CW00402577

Smoking Beauties

SMOKING BEAUTIES

STEAM ENGINES OF THE WORLD

ASHWANI LOHANI

Foreword by

MARK TULLY OBE

wisdom tree
C-209/1, Mayapuri, Phase-II, New Delhi-110 001
Ph: 28111720, 28114437

ISBN 81-86685-64-0

Editor : **Alka Raghuvanshi**

Copy Editor : **Manju Gupta**

Photo credits: **Phal S Girota, Paul Whittle, Ashwani Lohani, National Rail Museum.**

Layout and design : **Kamal P Jammual**

Printed at Print Perfect, New Delhi 110 064
Published by Wisdom Tree
C-209/1, Mayapuri, Phase-II, New Delhi-110 064

Foreword

The steam engine, which brought us the railway – revolutionising the way we travelled and covered the distances, opening up remote hamlets to the world beyond the 4-mile circumference that had restricted the people's lives, creating the great cities of the 19th century, heralding the age of mass production and distribution – replaced the handsome horse, the camel looking down his long neck, the steady, plodding bullock, and the humble donkey. In short, machines made by man from metals replaced flesh and blood, the living animals created by God.

Yet, when the first American steam engines huffed and puffed across the plains of Nebraska and Wyoming, the American Indians thought they were seeing iron horses. They were not the only ones to refuse to confine the steam engine to the status of a mere machine. In his anthology of railway journeys, the British writer and broadcaster, Ludovic Kennedy, wrote of others who, "thought of the night". That surely is why the steam locomotive is still known as the 'iron horse' and in India also as the 'black beauty'.

She, (we can't call a steam locomotive engine 'it') is living in a manner no other machine is, and is, of course, very beautiful too. But she's no slim, narrow-waisted, modern beauty; she has a might and a majesty which was described by the poet W.H. Auden in the commentary he wrote for a film made by the British Post Office about the Night Mail.

"Pulling up Beattock, a steady climb,
The gradient's against her but she's on time,
Past cotton grass and moorland boulder,
Shovelling white steam over her shoulder,
Snorting noisily, she passes
Silent miles of wind-bent grasses."

In *Smoking Beauties* Ashwani describes the start of his love affair with this beautiful lady at Cawnpore station. My affair started in India too, on Howrah station. The night train for Puri left well beyond the usual bedtime enforced in the 30s, years ago when children were still subjected to severe discipline. That added to the excitement of the vast forecourt crowded with anxious passengers trying to keep up with porters carrying incredible amounts of luggage. The unmistakable smell of steam in the air, the shrill whistle as the fussy little shunting engines brought the great expresses into the platforms, and the laboured breathing of the titans of the track starting on their journeys through the night added to the enchantment. We as children were not allowed to walk up the long platform to see the engine which was to pull our train. We were hustled into our compartment to ensure that the train didn't leave without us. But when we reached Puri the next morning, we were allowed to admire our engine, hissing contentedly at the end of her journey, and to sometimes even have a word with the driver.

Less happy were my later experiences with steam at Sealdah station where, struggling to hold back my tears, I would board the Darjeeling Mail to set off for the nine-month term at the New School, a night and a day's journey from home in Calcutta. But by the time we steamed into Siliguri, my life had changed gear, and I was back in the warm comradeship which only those who have been to a boarding school know. Boasting was one of the cardinal sins of old-fashioned schools, so I was careful not to claim credit for having a father who was a Director of the narrow-gauge railway which snaked slowly up the mountains to Darjeeling. The Darjeeling Himalayan Railway, or DHR as it's always known, laid on a school special for the second half of our journey. John Lethbridge has written a history of the New School which was founded as a temporary war-time expedient for the children of the Calcutta sahibs, like my father, who couldn't be sent back to boarding schools in Britain because the services by P and O and other shipping companies had been discontinued. In that history more than one pupil romances about the DHR and the small-in-size-but-large-in-heart locomotives which pulled our train. One has written of "... the toy train waiting for us in the siding at Siliguri — the air getting cooler as we chugged up the mountain, the pink cheeks of the ragged hill children as they ran along, laughing and shouting '*buckshees*' by the side of the track, the stop at Kurseong for hot soup of Bovril, and a quick visit to the little chemist just up the road to see their prize exhibit — a grisly two-headed lamb preserved in a large bottle of formaldehyde." I don't remember that ghoulish interlude in our journey, but I do remember the experience of another pupil who appears in the New School history — the experience of hopping off the train when the line looped and zig-zagged, running up the cud and clambering aboard again, as the gallant little engine returned to a more conventional way of climbing the mountain. That became more hazardous when the railway management decided to remove the steps outside the carriages to prevent ticketless travellers from hopping aboard.

My love of steam inspired by Howrah and Puri, Siliguri and Darjeeling, continued when I returned to Britain and went to school behind the engines of the London Midland and Scottish Railway, capable of speeds over 100 miles an hour, rather faster than the DHR. But steam is not all about speed. It's about living engines, their majesty, their beauty, and their large, large hearts.

Will the time come when any motor car, bus, or truck, demonstrates the courage of the DHR engines which at the age of well over 100 are still climbing the gruelling route to Darjeeling? If any motor vehicle does reach that great age, it will be a museum piece. The DHR locomotives are certainly not museum pieces; they are still in regular service. Indian Railways are to be congratulated for this achievement. In these days when everything is measured in rupees, in profit and loss, in bottom lines what other railway would recognise the importance of keeping a loss-making venture like the DHR going? It was an important railway when it was first built to open Darjeeling to the plains and to carry the 'champagne of tea' to markets in India and abroad; it performed an important function in my life, taking me to school; and it's now an important part of India's heritage which UNESCO has recognised by declaring it a World Heritage Monument. That's why I don't like the title toy railway. But running the DHR is not the only contribution Indian Railways are making to steam. They are building new engines and restoring old ones, they have revived a steam shed, and they run the world's oldest operational steam engine — the *Fairy Queen*. All this and much else, Indian Railways do as part of their deliberate policy to keep black beauties in steam, to demonstrate their commitment to the heritage they have inherited.

In Britain steam services are maintained by private railways largely manned by volunteers, and running on lines the mainline railways have closed because they are not profitable. The mainline railway companies often obstruct these ventures. I came across one private line which had been denied access to the mainline junction because it was thought that steam didn't fit in with the progressive image the railways were trying to portray — the age of high-speed electric and diesel engines. Steam engines are still occasionally allowed on Britain's main lines but always with reluctance. Yet, these private railways earn millions of pounds every year because the romance of steam lives on, as I am sure it will do in India. Ashwani's book is part of a growing movement to ensure that the Indian iron horses he loves, and I love too, will be alive to inspire that love in his grandchildren and great grandchildren.

Mark Tully OBE

Preface

My active involvement in matters relating to steam took shape during my tenure as Director of the country's only Railway Museum in New Delhi. Encircled by an assortment of black beauties, some of which are still in a smoking fit condition, I spent the best five years of my life in the sprawling museum, located in Diplomatic Enclave of Chanakyapuri. It was here that I learnt all that was to learn about these black beauties and the wonderful men who once maintained and ran them. It was here that I realised the positive difference that tempering by steam brings about in men. My subsequent stint in the portals of the Ministry of Tourism made me alive to the tremendous tourist potential that these black beauties possessed in plenty. I also became an active witness and partner to the Indian steam revival stories, besides playing the role of an active spectator to the global revival phenomenon. Awareness of the universal love, which the steam locomotives commandeered, ignited sufficient passion in me. It was then that I decided to pen down some stray thoughts, acquired knowledge and experiences of steam locomotives, more for the layman than for the steam expert. And that is how this book came about.

My better half, Arunima, who once got terribly annoyed with me for my over-indulgence with a stunning beauty by the name of *Fairy Queen,* has been the source of inspiration behind this book. Phal Ghirota, a good old friend, whose contribution to tourism photography is immense, has provided me with the bulk of the photographs for this book. Without his keen involvement and solid support, completing this book would have been impossible. My former colleague, Shalini Dewan who ensured smooth sailing of this book from concept to completion. I am also thankful to another close friend, Paul Whittle, one of the pioneers of the steam movement in the UK for his generous contribution of photographs, especially for the international section. His unstinted support and also the encouragement and guidance provided by my good friend, David Barrie from England, is unforgettable. Constant encouragement from Romesh Chandra Sethi, my revered senior and President of the Indian Steam Railway Society, Tarun Vijay, editor of *Panchjanya* and Livleen Sharma, a wonderful friend and my editor Alka Raghuvanshi who equally shared my love for the manuscript and the subject.

INTRODUCTION

INTERNATIONAL STEAM SCENARIO

INDIAN SCENARIO

Indian Hill Railways

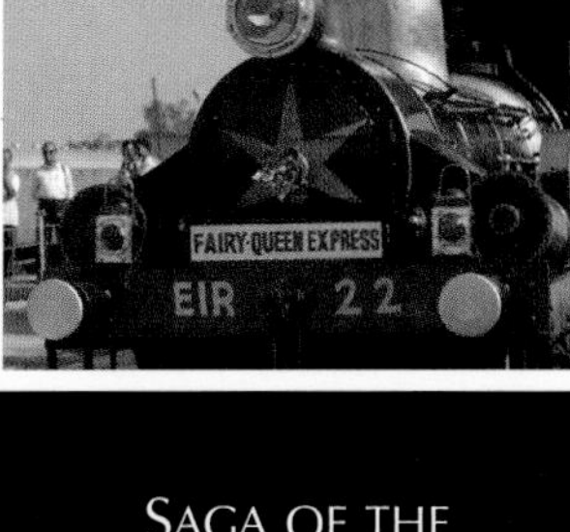

Saga of the Fairy Queen

Indian steam heritage preservation efforts

Introduction

My affair with steam began at the Cawnpore railway station in the early sixties, when, as an infant, the high point of any train ride used to be an escorted visit to the front of the train, where stood a big black steam locomotive, with its shrill whistle and the intermittent hiss of steam. This childhood romance was consummated much later in life, when, as the curator of the country's only National Rail Museum at New Delhi, I attained *nirvana* in the midst of a large collection of black beauties. Regular steamy interactions, slowly but steadily, initiated me into the niche global tribe of steam lovers. Since then I have firmly believed that this world is infested with two kinds of people — one who are the steamy types and the other who are not. Regular earthly interactions thereafter also convinced me of the unquestionable superiority of the former in all walks of life. This superiority, I believe, is fuelled in essence by the intense passion, which the steamy types have in plenty. Well, there may be other schools of thought, but I do not rate them worthy of much consideration, as I am too inclined to err on the side of steam!

Previous page: WP steam locomotive — poetry in motion.

Below : An early steam train of Great Indian Peninsular Railway.

What is so unique about steam locomotives, which evokes so much love and passion in the steamy types? What is unique to them, and not present in other contraptions that also move? Well, the presence of raw fire, that fires raw power in the belly of steam locomotives, is the draw. The unique sound which they make while moving, their rocking gait, their shrill whistle, their throbbing body and a very open design bordering on nudity are features that give a unique and attractive personality to these black beauties. The die-hard steam enthusiast firmly believes that a steam locomotive has an individual and unique personality of her own and like any beautiful lady, she requires the careful hands of a lover to stoke the fires burning within. It is precisely for this reason that in the good old days of steam, each locomotive used to be assigned to a crew of three – one driver and two firemen, by name. The men thoroughly knew their lady and obviously rode her well!

I would attempt an explanation once again, though on a relatively serious note. The invention of the steam locomotive has been the single most important event in the history of mankind. This invention marked the beginning of 'travel' beyond the frontiers. And once man began to travel, his vistas broadened and his thinking changed; in effect, his entire life changed. With the advent of the steam locomotive and its massive pulling power, it was but natural for railway systems to start growing around them. And they grew real fast. By the end of the 19th century, one could see steam locomotives chugging across national boundaries and traversing distances earlier considered impossible. Their numbers and spread grew so fast that people started identifying them with travel. And when they started getting replaced in favour of the more powerful diesel and electric locomotives, nostalgia began to surface.

The steam engine of James Watt fired the industrial revolution in the early 19th century and revolutionised travel. It was the first major step towards making the world a smaller place. It widened the horizons of the people by giving them the liberty and also the means to move beyond the confines of their villages and towns within a time-frame on a more economical scale. It enhanced trade and commerce, thereby bringing prosperity to the people and the society. In effect, the steam locomotive has been the harbinger of one of the major positive changes in societies, cutting across national boundaries. Unfortunately, the

introduction of diesel and electric locomotives in early 20th century led to depletion of the steam locomotive, both in importance and also in numbers.

Steam locomotives slowly started leaving the main lines and the branch lines thereafter. This trend became first visible in the developed economies and was followed after a certain time-interval in other underdeveloped and developing nations. The steam locomotive's departure led to nostalgia, as people began to miss it in many countries, particularly in the United Kingdom (UK), which, being the birthplace of the railways, obviously experienced the heart-rending pangs of separation. Love for steam, which surfaced, soon started getting expressed in a variety of ways. People formed clubs and steam societies to express their affection, to learn and relive the glorious past and also to experience live steam – though on a more limited scale. This marked the beginning of a global movement.

While James Watt with his tea-kettle experiments displayed the potential of the power of vaporised water as early as late 18th century, it was not until 1804 that Trevethick, the creator of the first steam locomotive in the world, proved with his early experiments at Coalbrookdale, and later at Penydarren in Great Britain, that a steam locomotive could pull a heavier load than a horse. These experiments, which gave the intellectuals an inkling of the times to come, however, could not lead anyone to visualise the extent to which rail transportation would develop in the future. In 1812, the first commercially successful locomotives were built by Mathew Murray to work for Middleton Railway near Leeds. They were able to haul 100 tonnes at 4 mph; they even replaced 50 horses and were the cause of putting 200 men out of work. William Hedley, the manager of Wylam Colliery, in 1813, built the *Puffing Billy,* an adhesion-type locomotive, which now holds the distinction of being the oldest steam locomotive in existence in the world today. The *Blucher,* built in 1814, was the first locomotive built by George Stephenson, who built a dozen locomotives and was established as the leading railway engineer by the time the Stockton and Darlington Railway was proposed. Locomotives built by Stephenson did the honours at the opening of this line in 1825.

The first quarter of the 19th century witnessed the emergence of the steam engine as a phenomenon of tremendous significance for mankind. The steam

locomotive not only unleashed the phenomenal power of steam, but also fuelled the development of industrialisation worldwide. This resulted in the growth of a technology, the role of which in the emergence of the 19th century civilisation can only be described as 'stupendous'. The steam locomotive came to be recognised as the harbinger of the industrial revolution. The period between 1804 to 1825 witnessed scenes of intense activity. Steam locomotive technology was experimented with at various places in the UK, especially in the collieries. George Stephenson played a pioneering role in building the *Blucher,* a highly successful steam locomotive in the year 1814. He was also instrumental in taking the first step towards building a mainline steam railway in 1821, when he was appointed the locomotive engineer at the Stockton and Darlington Railway. This line, opened with great fanfare in 1825, marking the birth of the first public steam railway of the world.

The intense passion which the steam locomotives commandeered from day one, was amply reflected in 1829, when only four years after the advent of the

Steam train at a dockyard.

219
WP

railways, the world witnessed their first race. The Rainhill trials were held before the opening of the Liverpool and Manchester Railway, in which Stephenson's *Rocket* was a successful entry. This legendary locomotive is still preserved with pride at the London Science Museum. Its exact replica, built in 1974 by Mike Satow, one of the pioneers in preservation of railway heritage, enthrals visitors at the National Railway Museum, York, by taking them for short joy rides.

Soon the railways started making their presence felt all over the globe. In the next 25 years or so, railways made inroads in England, France, Germany, Holland, Italy, Spain, Russia, and the United States of America (USA). India too joined the bandwagon in 1853. The plans started with a controversy over the wisdom of starting a railway in a backward country like India. It was argued that how could the destitute Indian, who did not possess even an anna (about 20 paise), be persuaded to pay train fares in preference to jogging peacefully on his bullock cart? After all, India was not a flat country like Russia, nor a small one like England. Some areas were inaccessible by vehicles of any type; some were in dense forests, infested with wild animals, vermin and malarial insects. Others did not receive an inch of rainfall in a year. In addition, the basic equipment and facilities required for such large-scale construction work were minimal.

Yet, the railways had to come to India.

Construction work on the Ganga Canal began in 1845, where it crossed the 2.5 miles-wide Solani Valley (90 miles north-east of Delhi) on a 15-arch aqueduct with long approach embankments; the earth for the latter being transported along light railway lines. The ironwork for the wagons came from England and they were built to 4ft 8½ ins. gauge, being moved at first by hand and later by horses. On 22 December, 1851, a locomotive started work – the first in India! Named *Thomason,* it worked for a few months before it met with an accident. The *Report on the Ganges Canal Works* stated: "The water had been drawn off, and it was supposed that the fire had been entirely extinguished. A storm with wind, brought the fire and fuel which were in the furnace, into action, and destroyed the casing together with a number of tubes, placing the locomotive completely out of use."

Thomason was a six-wheeled tank locomotive, an E.B.Wilson 2-2-2 WT. E.B.Wilson stands for the name of the manufacturer; 2-2-2 for the wheel arrangement of the locomotive; W for broad gauge (5ft 6ins.) and T for tank

Facing page: Raw power — WP steam locomotive crossing a bridge.

(which stores water mounted on the engine) type locomotive. At this stage, it is important to clarify to the lay reader that a steam locomotive is generally identified by its wheel arrangement. It normally has three sets of wheels: front, driving and trailing. A 2-2-2 arrangement would signify two front, two driving and two trailing wheels. On an axle, there are two wheels. Most of the earlier steam locomotives were of the six-wheeled type with 2-4-0 and 0-4-2 wheel arrangements. The wheels are powered or rotated by means of a drive generated by the movement of pistons. (Talking of pistons, there are two cylinders, inside which the pistons powered by pressurised steam from the boiler move the 'driving' wheels on either side). These driving wheels, unless there is only one, are coupled by using coupling rods for improving the adhesion. There are two types of cylinder arrangements — one in which the cylinders are inside the main frame of the locomotive and therefore hidden from public eye, and the other, obviously has cylinders outside the frame. In addition, there are valve gears, which are mechanical arrangements for timing the injection of steam into the cylinders for driving the pistons. These are also identified by their designs like *Walschaert, Caprotti,* etc. A tender is a wheeled attachment to the locomotive and it carries water and coal. Locomotives without the tender have water tanks mounted on the sides of the boiler and are called tank-type locomotives. The axle load is the weight of the locomotive in tonnes divided by the number of axles. The various classes of steam locomotives are identified by alphabetical prefixes before the unique number of the locomotive. These prefixes like WM, QA, QB, ZB, ZF, ZF1 YM, X, F are used to signify what these mean in terms of the name of the manufacturer, boiler pressure, wheel arrangement, cylinder arrangement.

In November 1850, the East India Company decided, on the recommendations of the Indian Government which was worried about the stability of trains in cyclonic winds, that the gauge in India should be 5ft 6ins. The second steam locomotive to reach India, the *Falkland,* complied with this ruling. Brought by the contractors, Favell & Fowler and built by the Vulcan Foundry, this locomotive commenced running at Bombay on 23 February, 1852. This was again an E.B.Wilson, four-wheeled tank locomotive. On 16 April, 1853, the railways made their formal beginning in India when a 14-coach train, carrying 400 elite members of the society and hauled by the historic trio of steam locomotives, *Sultan, Sahib* and the *Sindh,* travelled from Boribunder to Thane, undertaking a

Facing page: The fully decked up majestic Canadian Pacific WP steam locomotive.

7585

journey of 34 miles on the Great Indian Peninsular Railway. The train steamed out to the accompaniment of the Governor's band and amidst the applause of a vast multitude and the booming of a 21-gun salute. The day was observed as a public holiday.

It is of essence that the saga of the railways is not a chicken and egg story. Steam locomotives came first and the railway systems grew around them, slowly in the initial years and rapidly thereafter. The locomotive foreman, the person who commandeered and maintained steam locomotives, was indeed the most important person in the railways, anywhere and everywhere. Steady improvements in the steam locomotive technology kept pace with the growth of the railway systems till about late 1950s and early 60s, when the Western world started discarding them in favour of the more powerful diesel and electric locomotives, mainly for operating considerations on account of much enhanced passenger and freight requirements. India too followed suit. Much later, large-scale destruction of steam locomotives followed in right earnest, particularly in the late 1980s and early 1990s. December 1995 saw *WL 15005* hauling the last steam train on the broad-gauge system between Jalandhar and Ferozepur. Finally in February 2000, with the closure of the steam shed at Wankaner, steam bid adieu to the mainline metre-gauge network in India. Meanwhile, steam had also started giving way to the narrow-gauge network and presently only two lines — the Darjeeling Himalayan Railway on the North Frontier Railway and the Nilgiri Mountain Railway on the Southern Railway — are the last remaining vestiges of a glorious era of steam for the Indian Railways.

Steam was considered good for the railways so long it was possibly cheaper to run steam locomotives in comparison to other forms of motive power, namely diesel and the electric. However, there can be no doubt that steam will always continue to have a strong place in the tourism sector for the prime reason that it evokes nostalgia for rail travel of a bygone era. It is steam locomotives, which laid the solid foundation for today's modern railway system across the globe.

Next page: Crossing the Deccan is a YG-class metre-gauge steam locomotive at the head of the Kachiguda-Ajmer Express.

And this is precisely why the world is now witnessing a return of the steam era in the developed as well as developing economies. Steam is making a serious comeback and a successful one at that. The return of the steam locomotive in a limited way, both for heritage and tourism,warrants some good times ahead for

the inveterate traveller and lover of railways in both the developed as well as developing countries.

Below : Brahmaputra by steam a tourism package of NF Railway.

International Steam Scenario

The current international steam scenario warms the heart of steam enthusiasts around the world. Encouraged by the positive tourist response, fuelled out of passion and love for steam engines, state railways, a number of private societies and many individuals worldwide have decided to awaken some of the smoking beauties.

Baldwin 2-8-0 No 1578 built in 1919 heads a train of empty sugar cane wagons on the narrow gauge Rafael Freyre system in Holguin Province in the east of Cuba.

The more than 37,000 steam locomotives which ran on class 1 railroads in the USA at the end of World War II, were reduced to around 17,00 in 1958, and reached negligible proportions by the late 60s. Steam also departed from the national railways of Holland in 1958, Belgium in 1966, UK in 1968, Norway in 1970, Denmark in 1971 and France and Japan in 1975. The subsequent period, however, saw substantial work being done in various countries towards live

preservation of steam locomotives. This preservation activity can be placed in three categories : railways run regularly by preservation organisations, retained or repurchased steam locomotives by railways to haul tourist trains, and locomotives owned by preservation organisations run on railways, which otherwise provide a normal transport service. Live preservation of steam locomotives has generally been very active in the UK, countries of north-western Europe and in English-speaking countries like the USA, Canada, Australia, New Zealand, etc. In these countries the steam locomotives diminished with speed, fuelled by the intense competition offered by road and air travel. Countries of the erstwhile Communist bloc and those in Latin America, Africa and Asia, with the exception of Japan, however, do not have any appreciable live steam locomotives.

Fire through the snow.

The pioneer in this segment has been the Talyllyn Railway Preservation Society, which was set up in the UK in 1950. This 7-mile long railway in North Wales is unique in that it still has some steam locomotives and coaches dating back to the 1860s. Another remarkable railway is the *Festiniog* that is about 30 miles from Tallylyn, also in the Wales. Originally built in 1836, to carry slate from quarries to the port, the trains were powered by horses or gravity till 1860. This railway is by far the biggest narrow-gauge preserved steam operation in the world. The Festiniog Railway Society has its origins in a public meeting held in 1951, after which the restoration work commenced and the complete line reopened in 1977. This 597-mm gauge railway, running from Blaneau Festiniog to Porthmadog, is about 14 miles in length and is one of the biggest success stories in the world in the field of steam heritage tourism. The double-fairle class of steam locomotives which run on this line are real beauties.

The Boston Lodge Works on this railway is also a unique facility as it utilises a bare minimum infrastructure and negligible staff; it is manufacturing new while restoring old steam locomotives, besides carrying out running maintenance of the trains operating on the railway. This workshop is a classic example of how something considered impossible by conventional standards can be achieved with grit and determination. The heart-warming examples set by the Talyllyn and the Festiniog Railways have been widely followed in other parts of the UK and by other countries. The first steam-worked tourist railway in France was the brainchild of some railway enthusiasts, notably Jean Arrivez, who regretted the decline and disappearance of light railways, once very common to that country. The Chemin de fer Touristique de Meyzieu, one-mile long, 600-mm gauge line operated by enthusiasts, and located on roadside east of Lyons, started in 1962. The line opened in 1962 and was a tremendous success. Many other steam tourist railways were thereafter established in France, notable amongst them being the 700-mm gauge Abreshviller Forest Railway and the Chemin de Fer Touristique des Landes des Gascogne, a standard- gauge line in the south-west of France. In Belgium, preserved steam operations started in 1966 on a 7-mile long metre-gauge line in a scenic part of the Ardennes, operated by the Tramway Touristique de l'Aisne, an organisation set up by an association of light railway enthusiasts. Holland has the Tramway Society operating steam-hauled tourist trains. Sweden, Switzerland, Austria, Hungary, Spain, Czech Republic and

Facing page: Steam train on the Tallelyn Railway in Wales.

LATVIJAS DZELZCEĻŠ
L-5225

Germany also have preserved steam operations. In the USA, steam railways have become more closely involved with the tourism industry than elsewhere, and preservation of heritage is confined mainly to the railway museums. One of the most tourist-oriented preserved steam operations can be experienced at Tweetsie Railroad, near Blowing Rock in North Carolina. Another world-famous steam operation in the USA is the *Silverton* train. This runs on a 48-mile long, 915-mm gauge line, built in 1882, from Durango to Silverton and hauled by large 2-8-2 steam locomotives. Australia has the famous *Puffing Billy* train that runs on a 9-mile long, 762-mm gauge branch line of the Victorian Railway from Belgrave to Lakeside. This line, which opened in 1900, passing through a picturesque route, was always very popular with enthusiasts. The *Puffing Billy* Preservation Society was formed in 1954 with the aim to reopen the line from Belgrave to Lakeside, earlier closed due to a landslide. The *Puffing Billy* locomotives are a distinctive class of 2-6-2 tank, Australian-built, but based on a Baldwin design. As with the earlier British Railway preservation schemes, the establishment of many other railway preservation groups around the world has followed the example of the *Puffing Billy* in Australia.

Among the various voluntary organisations that were involved in the operation of steam locomotives, the one with the most extensive operations has been the Eurovapor, the European Association of Friends of Railways for the maintenance of steam locomotives. This organisation, based in Zurich, was founded in 1962. It has 1,000 members and operates steam trains over five routes in three different countries. Its busiest service is on the 23-mile long, 760-mm gauge Austrian Federal Railways branch from Bregenz to Bezau. Its other main services are on Waldenburgerbahn and the metre-gauge lines between Worblaufen and Solothorn, between Worblaufen and Worb in Switzerland, and on the 9-mile long standard gauge line from Haltingen to Kandern in Germany.

India is one of the late entrants on the steam scene for the simple reason that it is a country that bid farewell to regular steam operations rather recently – not because of any love for steam locomotives, but because the diesel and electric replacements were slow to come by. But again, India's biggest folly has been that almost the entire steam set-up was destroyed and was not gradually phased out. Steam locomotive graveyards became a common sight from the late 1970s to the early 90s and some misplaced railway officers could be witnessed proudly laying

Facing page : The only working standard gauge steam locomotive in Latvia, this ex-Russian L Class 2-10-0 freight locomotive built in 1952 is seen on a special passenger train at Riga Central Station.

MI · 631

emphasis on the numbers destroyed and taking credit for the destruction. Steam locomotives were pulled apart and sold as scrap, without realising the irreparable loss of a heritage that this exercise caused the nation. The developed economies have, however, been smarter, with the result that they still have a substantial fleet even after steam left the main lines. In stark contrast with India, however, is the UK, where steam operations at present are still substantial and steam heritage is a very important component of the tourism scene. Similar is the case with many other countries, with the exception, of course, of China and Myanmar where steam is still doing its regular rounds on the main lines. It has, therefore, been possible, when the time came, for the lure of the steam to surface far and wide, spreading across continents. One can now come across working steam locomotives in many countries in Africa, America, Asia and Europe. While China and Myanmar still have the mainline steam locomotives, Angola, Botswana, Sudan, Argentina, Brazil, Chile, USA, UK, Germany, France, Poland, Romania, Indonesia, Sri Lanka and Pakistan, to name a few, are running steam heritage tourist excursion trains. The *Fairy Queen* and the *Royal Orient* in India, the *Blue Train* in South Africa and the *Flying Scotsman* in the UK have now become household names amongst the prestigious and luxurious steam-hauled tourist trains on the planet Earth.

A steam enthusiast is a passionate tourist and he will travel far to experience the object of his passion – the black beauty. To meet the needs of such a tourist, whose idea of a holiday is a ride on or behind a black beauty, and for which he is prepared to globe-trot to any distant corner of the world, an exclusive tribe of steam tour operators has emerged. Steam tours to Indonesia, Cuba, Myanmar, USA, UK and India are now becoming increasingly popular. The glorious era is also being relived in numerous railway museums around the world, where steam locomotives generally form the core theme. The National Railway Museum, York, National Rail Museum, New Delhi, Railway Museum at Nurenburg and many other such museums have exquisite collections of one of the most beautiful and popular steam locomotives, which once ruled the iron rail. Many railway museums are also maintaining steam locomotives in working order for hauling tourist trains. Substantial work in this direction is also being carried out in South Africa. Ushuaia in Argentina has emerged as a centre for research on steam locomotive technology.

Facing page : Another view of the Tallelyn railway in Wales.

The large and widespread resurgence of steam locomotives around the world is a pointer to the interest which has resurfaced after these beauties were taken off regular services. Thus, I would call this a movement primarily fuelled by nostalgia and a desire to preserve for posterity, what is rightly responsible for rapid industrialisation the world over in the 19th century. UK, which in 1825 heralded the advent of railways, has again emerged as a leader in the comeback effort of steam locomotives. The Heritage Steam Railway in the UK, which started in late 1950s, is now no longer a movement of only dedicated railway enthusiasts, but has evolved into a big and growing industry to become a key factor in tourism programmes in many areas. From the great little trains of Wales which feed hundreds and thousands of visitors to the busy commuter lines, such as Dartmouth Railways, the heritage railways in the UK are thriving and thrusting in

A fine collection of preserved steam locomotives, seen here at the steam roundhouse depot at Luzna-Lisany near Prague.

all directions. As a result, 108 operating heritage railways and 60 steam centres have evolved at present. These railways with a total mileage of 427 and 570 stations have a length greater than the London underground. Preservation of various new heritage railways and expansion of the existing railways is on the cards and the total route mileage is likely to cross the 600 mark in the near future.

While discussing the development of steam heritage tourism with the mandarins of the various Bhawans of the capital of India, I quite often, finding no other means to put forth a convincing argument, am forced to point at global statistics, which are really astounding. When the mandarins display their misplaced concern about environment, I have to draw their attention to the much more stringent environmental standards in the UK, which is permitting the operation of

A 1' gauge working replica of the B class steam locomotive of the Darjeeling Himalayan Railway seen at the Fairbourne railway in Wales.

Stepney an A1X Class 0-6-0 tank locomotive built in 1875 for the London, Brighton and South Coast Railway seen at Sheffield Park locomotive depot on the Bluebell Railway. The livery is what Stroudley, the designer called his 'Improved Engine Green'—but the story is that he was colour blind and none of his staff were brave enough to tell him of his mistake!

so many steam engines. UK alone has a heritage railway network, the magnitude of which is really astounding. With over 1,300 operational steam locomotives at present, out of which around 500 are working on main line of standard gauge, 545 working on heritage lines of standard gauge and 250 working on the narrow-gauge lines, the heritage railway network in the UK is possibly the biggest success story of its kind in the world and easily the most convincing tool for propagating steam tourism.

I am convinced of the need to further the cause of steam internationally. I am equally convinced of the ability of the international steam community of being able to do so. Yes, getting together will help. Getting together on an international

platform that provides mutual encouragement, mutual awareness and mutual exchange is the need of the hour. Steam locomotive exchange programmes between countries operating steam trains in one way or the other is necessary. Possibly one such exchange can be between the Darjeeling Himalayan Railway in India and the Festiniog or the Leighton Buzzard Railway in the UK. There can be many more and much varied exchanges, contributing in no small measure to the cause of steam. One such platform fuelled by the strong desire to come together emerged shortly during the first ever International Steam Trains Congress, held in Switzerland in October 2003. At the conference, it was also decided to form an international organisation and seek UNESCO affiliation.

Next page : Class 4 2-6-4 tank locomotive No 80151 stands at Horsted Keynes Station with a demonstration freight train on the Bluebell Railway in Sussex.

Below : Baldwin 2-8-0 No 1578 built in 1919 seen at Ecuador Mill. The steam locomotives run on low-grade oil from Cuba's oilfields at Varadero.

80151
75A
SC
KEYNES

Indian Scenario

Broad gauge: 22 December, 1851 marked the beginning of the steam era for the Indian Railways when *Thomason,* a 4ft 81/2 in.-gauge locomotive started work during the construction of the Solani aqueduct, 30 miles north-east of Delhi. It was a six-wheeled tank locomotive, an *E.B. Wilson WT.* The second steam locomotive to arrive in India was the *Falkland,* a broad-gauge 5ft 6in.-steam locomotive which commenced running at construction sites near Bombay on 23 February, 1852. Then came *Sultan, Sahib* and *Sindh,* the three historic locomotives, which hauled the first train on the Indian soil from Boribunder to Thane, a distance of 21 miles on 16 April, 1853. The sixth in line was the *Fairy Queen* of 1855 vintage, which still hauls tourist trains and has now come to be regarded as a national treasure.

Previous page : YG-class metre-gauge steam locomotive hauling a train on the Northern Railway.

Below : A fully decorated WP steam locomotive at a wayside station.

While the earlier locomotives were of the inside-cylinder type, in 1855, however, Kitson Thompson and Hewitson built a few outside-cylindered locomotives for the Madras Railways and the East Indian Railway and one of these locomotives, the *Fairy Queen,* is still preserved in working order in New Delhi. In 1877, a 4-4-0 mixed traffic locomotive with outside cylinders was introduced in the North-Western Railway and in 1879, the traditional British passenger, four-coupled design appeared on the Indian scene, on the Great Indian Peninsular Railway. This design continued as the mainstay of passenger services all over India, with the goods services remaining with the contemporary 0-6-0 locomotive. The main exception to this trend was a series of interesting small-wheeled 4-6-0 locomotives with outside cylinders specially introduced in 1880, to cope with the gradients on state lines; these locomotives were the L class on the North-Western Railway.

With the induction of numerous types of locomotives in service towards the end of the 19th century, a need was felt to go in for standardisation in the designs of steam locomotives. A committee was appointed to prepare designs of a limited number of standard classes in the hope that the future requirements of various railways would be met accordingly. The first report of this committee in 1905 specified a 4-4-0 for passenger and a 0-6-0 for goods with 180 psi or working pressure, a Belpaire firebox, inside cylinders, balanced slide valves and Stephenson link motion. Later in 1910, an alternative light boiler was also specified for lines requiring more power. In 1906, with the revision of permissible standard dimensions, more broad-gauge locomotives were designed – a 4-6-0 and 4-4-2 for passenger, and 2-8-0 and 2-6-4 for goods services. These standard types were known as BESA classes and were built in large numbers. Superheating was subsequently introduced in 1912, for extracting more power from steam locomotives.

After World War I, the demand for locomotives increased due to a large upsurge in traffic. Accordingly, a Locomotive Standards Committee was set up in 1924 to update and re-standardise the existing BESA designs and to recommend new types of more powerful locomotives. For the broad gauge (5ft 6ins.), three types of passenger 4-6-2 and two types of freight 2-8-2 locomotives were recommended and these constituted the various X classes of locomotives. This period also saw the evolution of huge Garrett locomotives and the deGlehn

WP

compound 4-6-2 on the Bengal-Nagpur Railway. In 1943, a new series of modern tank locomotives WM (2-6-4 T), WO (2-4-2 T), WV (2-6-2 T) and WW (0-6-2 T) were introduced, with only WM being built in large numbers. During World War II, there was a heavy increase of traffic on the Indian Railways and imports were resorted to from the USA and Canada. These consisted mainly of AWD (American War Design) and CWD (Canadian War Design) of locomotives. After the war, WP, a 4-6-2 type of passenger locomotive was introduced after years of research on boiler efficiency, valve gear and rail wheel interaction. This had a maximum axle load of 181/2 tonnes and ultimately 755 such locomotives were put on line. This photogenic bullet-nosed locomotive came to be identified as the symbol of Indian Railways. Subsequently, WG, a 2-8-2-type locomotive, was also designed using the parts of WP, mainly for hauling goods trains. Altogether 2,450 locomotives of this type were put on line. These two designs continued to be the mainstay of the steam traction in India, till these were withdrawn from service in 1995.

Steam locomotive building activity in India began in the 1870s at the Ajmer Workshop of the erstwhile Bombay-Baroda and Central India Railway, where locomotives were assembled using the spare parts that came with new locomotives. Real manufacture, however, started with the setting up of the Locomotive Works at Chittaranjan in West Bengal, in 1950. Steam locomotives continued to be manufactured here till 1972. The production of BG steam locomotives dawned with the manufacture of *Antim Sitara,* a WG, in 1970, and that on the metre gauge with the manufacture of *YG 3573* in 1972. It is no coincidence that an eminent railwayman, Ramesh Chandra Sethi, who was midwife to the last metre-gauge locomotive at Chittaranjan, is now the active president of the Indian Steam Railway Society!

The metre-gauge railway in India received comparatively little publicity over the years. Lacking the glamour and importance of the broad-gauge lines, and without the individuality of the narrow-gauge railways, they carried out their vital role largely in oblivion. It is, however, noteworthy that at one time, the metre-gauge route was at approximately 17,000 miles, which was not very much lower than the broad gauge total of 20,167 miles. There were many large areas, particularly in the far north, the north-west and parts of the deep south, where the metre gauge reigned supreme. It was possible at one point of time to travel from a

Previous page : Activity at a steam locomotive maintenance shed on WPs.

southernmost station in India to the far north-east corner of Assam entirely on metre-gauge lines; it would take a long time, but it would have been a memorable experience! The metre-gauge system was predominantly single-track and the contrasts are remarkable, ranging from the Nilgiri rack railway with gradients of 1 in 12½ to monotonous and dusty level stretches in the northern plains. Although the metre-gauge rolling stock has always been more standardised than the individualistic broad- and narrow-gauge lines, there were many differences in detail and special designs, whose origins dated back to the era of the smaller companies. The ownership, financing and working of these various lines is an interesting story in itself.

The first broad, 5ft 6in.-gauge, trunk lines were completed in the 1860s. They were built by British companies, with a Government of India guarantee of interest on the capital, and initially proved a heavy burden on India's finances. It was then proposed that the State should build further lines to develop the system and an argument ensued as to whether these should be lightly-built broad-gauge railways or constructed for a narrower gauge. In the end, the latter course was adopted, except for some strategic lines the choice of the actual gauge was left, rather unkindly, to the Viceroy of that time, Lord Mayo. He based his decision on two assumptions: the overall width of a carriage to seat four on a side should not be less than 6ft 6ins. and the overall width should not be greater than twice the gauge. The British Government confirmed his choice of 3ft 3ins. but it so happened that a commission was then debating the introduction of the metric system in India and, in anticipation, the gauge was therefore altered slightly to more than a yard.

The early lines were comparatively short, but the system soon developed. The first section to be opened in 1873, was part of the Rajputana State Railway which ran from Delhi south-westwards to Rewari, serving enroute an important salt lake at Farukhnagar. This railway was extended to Bandikui to meet another line running westwards from Agra and then pushed forward to Ajmer, where in due course it was joined by the Holkar State Railway and its long extensions coming north from Khandwa (on the broad-gauge Great Indian Peninsular Railway). The amalgamated system was known as the Rajputana Malwa State Railway; meanwhile, the original line through Ajmer had reached Ahmedabad and had thus become part of an important new route from Delhi to Bombay. The initial

Next page : Full steam ahead : WP struggles to pull a heavy load.

5 WP

classes of metre-gauge steam locomotives were mainly of British origin, with the exception of a few German locomotives. With the increase in traffic towards the end of the 19th century, the need for more powerful locomotives was felt and three new 4-6-0 designs with inside frames and up-to-date proportions were therefore evolved.

These included the class A for passenger service built by Neilson and Co. for the BNR with inside valve gear and largest (5ft diameter) coupled wheels. Class B was built for mixed traffic, to haul both passenger and freight trains, again built by Neilson and Co. with outside Walschaerts valve gear and 4ft diameter-coupled wheels. A third design was the mixed traffic type, built by Sharp & Stewart for the Rohilkund and Kumaon line with Walschaerts gear and 4ft diameter-coupled wheels. Further in 1903, three BESA designs of steam locomotives evolved. These included the passenger design, 4-6-0 with Walschaerts valve gear and coupled wheels with a large boiler, the mixed traffic design with 4ft diameter wheels and most of the parts freely interchangeable with the passenger class and a 4-8-0 design with a still large boiler for heavy-duty freight operations.

These standard designs (mainly 4-6-0) were widely used on the metre-gauge network. There were, however, many modifications and various railways, while accepting the basic designs, finalised their own standard details to avoid multiplicity of spare parts. Local conditions, particularly the quality of water supply, affected the tender and boiler specifications. Hence the boiler specifications were further modified with the introduction of superheating in 1912. After World War I, a new series, known as the IRS classes, were evolved to cater to the increased traffic and cope with the inferior coal allotted to the railways that necessitated a wider firebox on the mainline locomotive.

During World War II, heavy traffic, particularly on the strategic Assam Railway could not be met satisfactorily and a number of 2-8-2 locomotives were obtained from America. Called MAWD, they were later changed to WD (War Department). Garatts were also used on the hill sections in Assam, but somehow these were not popular. Post-war designs, four in number, were built in large quantities to cope with the increased traffic and replace averaged locomotives. By 1972, there were 871 YP 4-6-2 T and 12 YM 2-6-4 T for long-distance passenger trains. A unique species is the large Class X Swiss locomotive, which

still works on the Nilgiri Mountain Railway, operating between Mettupalayam and Conoor. These locomotives, which were built during the period 1914–1922 to metric dimensions, are compound locomotives with all four cylinders outside the frames. On the easier section of the line, these work as two-cylinder simple locomotives. The two low-pressure cylinders driving the rack wheels are immediately above the high-pressure cylinders and the working of complicated Walschaerts-type valve gear will fascinate a steam enthusiast as well as an engineer. The rack system provides extra adhesion, which also prevents these locomotives from rolling down on steep inclines.

Narrow Gauge

The first narrow-gauge railway to use locomotives was a 20-mile line built by His

Mail train hauled by a WP at a station.

Highness the Gaekwar of Baroda, to connect the town of Dabhoi with the nearest point of the Bombay – Baroda and Central India Railway at Miyagam. The gauge was 2ft 6ins. and the first section was opened in 1862, this being originally operated by bullocks; in the following year, the *Illustrated London News* carried an engraving of the scene and commented: "The rude native cart, with its lazy bullocks and drowsy driver, are, we may now expect, soon to be replaced throughout this rich province by these busy little tramways."

All decked up for the high-paying tourist, the Royal Orient, being hauled by a YG steam locomotive.

The Gaekwar of Baroda had already ordered three small tank locomotives for the line and these were built by Neilson & Co. in 1863, but they proved too heavy for the light (13 lbs to a yard) rails. It was not until 10 years later, when the line was rebuilt with heavier (30 lb) rails that these locomotives could be used on a

regular basis. In later years, this system was greatly extended and included isolated sections in other parts of the Gaekwar's domains; nowadays, these lines form part of the Western Railway zone.

Next came the first section of the Darjeeling-Himalayan line, opened in 1880. As it followed the existing road as far as possible, the steep gradients and sharp curves persuaded the promoters to adopt a 2ft gauge; even so the original alignment proved too steep for the first locomotives and loops and Z-type reverses had to be introduced in certain places. The same gauge was used for the company's Teesta Valley line and even for the long, comparatively level extension south-westwards to Kishanganj, built in 1914-15. During 1881-94, nearly 11 further narrow-gauge lines (three at 2ft and the rest at 2ft 6ins.) were introduced in various parts of the country and for a variety of purposes, but none of these survived in their original forms. Some have quietly disappeared and metre-gauge lines have replaced the others. One deserves particular mention : in 1883 the Thakur Sahib of Morvi visited England and met a highly efficient salesman from Kerr, Stuart & Co. (who were then agents, not makers) with the result that the firm shipped out the equipment for a complete 2ft 6in.-gauge railway in Morvi state. In later years the main lines were converted to metre gauge, but some of the old material was used again to construct a network of light tramways in the Morvi area and one of these survived till very recently. Maharaja Scindia of Gwalior was another early customer of Kerr, Stuart & Co.; he started a private 2ft-gauge railway in his palace grounds in 1893 and from 1899 onwards, this was developed into a much larger public system, serving his widespread dominions.

Two important events occurred in 1897. In that year, the first sections of the 2ft-gauge Howrah-Amta and Howrah-Sheakhala Railways were opened, catering to local traffic in the area immediately west of Calcutta. With a guarantee from the local district boards, these lines were managed by T.A. Martin & Co. of Calcutta and several other lines were later sponsored by this firm, built at 2ft 6ins. gauge. The year 1897 also saw the opening of the first section of the 2ft 6in.-gauge Barsi Light Railway. Engineered by E.R.Calthrop (better known in England in connection with the Leek & Manifold line), this was a highly successful attempt to show the carrying capacity that could be achieved on a narrow-gauge railway, using high standard equipment. The large steam locomotives were an

improvement on earlier usages and paved the way for a general improvement in standards on the later narrow-gauge lines. Moreover, some uniformity was also achieved at last regarding the actual gauge to be used. As a few strategic lines were to be built in the North-West Frontier area, as additional stock could be required hurriedly in the event of an emergency; hence it was important to have an agreed standard. In 1897, a conference in India recommended the 2ft gauge, but in the following year the War Office advocated the 2ft 6in.-gauge for military requirements throughout the British Empire and which the Indian Government accepted as its future standard.

During the first decade of the 20th century, several important lines were opened, including the spectacular Kalka-Simla Mountain Railway, two strategic lines in the far north-west and the beginning of the extensive narrow-gauge system of the Bengal-Nagpur Railway. In the next 10 years, despite war-imposed restrictions, the total mileage doubled, and some further expansion took place in the 1920s despite serious competition offered by road traffic. At the beginning of 1940, there were over 4,000 miles of narrow-gauge railways in India. Many were privately owned or were the property of native states, but the major railway companies operated most of the larger systems. Ownership of the latter also varied considerably, thus the Great Indian Peninsular, North-Western and Eastern Bengal systems were State railways, i.e. owned and operated by the Government of India; the Bengal-Nagpur and Bombay, Baroda and Central India Railways were State-owned, but operated by London-based companies, and the Gaekwar's Baroda State Railway was a native state line. During the World War a few narrow-gauge lines were dismantled, but the majority enjoyed the general upsurge in traffic throughout India. Then, on the eve of 15 August, 1947, with the country's partition, some 490 miles of narrow-gauge track passed into the jurisdiction of Pakistan. By 1951, the Indian Government had taken over the remaining company-operated major railways and also the native state lines. The next step now was to regroup the railways into six, which later increased to nine and now very recently to 16 zonal systems. This left only a few narrow-gauge privately-owned concerns and of these, one (Barsi Light Railway) was purchased by the government in 1954 and one (Central Provinces Railways, consisting of lines from Ellichpur to Yeotmal and from Pulgaon to Arvi) was worked by the Central Railway anyway. The others, in common with the government-owned

narrow-gauge lines, found it increasingly difficult to cope with increasing competition from the road and rising costs.

By far the most artistic and beautiful locomotives have been those in use over the narrow-gauge network of the Indian Railways. The first narrow-gauge locomotives to be used in India were the 0-4-0 tank locomotives on the Gaekwar of Baroda line, in 1863. The 0-4-2 tender locomotive in 1891 and the Kitson design 0-6-2 followed these in 1912. The famous Darjeeling line started with eight 0-4-0 side-tank locomotives, which were not successful and new drawings had to be evolved. In 1889, the famous B class 0-4-0 tank design evolved and this being highly successful, is still the only type of steam locomotive in use on this section. The Darjeeling design was however not successful on the picturesque Kalka-Simla

WG class steam locomotive being coaled at Saharanpur.

Railway for which a large 2-6-2 T-type was evolved and this locomotive, after necessary improvements, continued as the standard motive power till the introduction of diesel locomotives in the year 2000. The hill railway in Matheran used 0-6-0 T-type locomotives with radial movement for outside axles. The small tank-type locomotive continued on the NG, till large 0-8-4 T (1877) and 4-8-4 T (1905) were introduced on the Barsi Light Railway. The Bengal-Nagpur Railway, South Indian Railway and the Great Indian Peninsular Railway for their narrow-gauge sections also adopted this design of eight- coupled type.

The tank-type locomotives had inherent limitations due to their restricted coal- and water-capacity and this led to evolution of a 4-6-2 tender design for passenger services in 1906. In 1915, superheated versions of these designs were introduced. After World War I, in 1925, a list of standard types for the 2ft-gauge was issued. All these types were superheated and had 160 psi operating steam pressure. Only the ZB, ZE and ZF were standardised, out of which ZB and ZE were ordered in substantial numbers. Five ZF 1 class 2-6-2 T locomotives were also obtained. These were similar to ZF with the exception of Walschaerts valve gear as against Caprotti valve gear used in the ZF class. In addition, two new designs and ZD were also introduced in all the batches. These had a higher boiler pressure of 200 psi. For the 2 ft-gauge, three classes were standardised, these being the QA, QB and QC. QA and QB were 2-6-2 with 3/4 and 6 T axle loads while QC was 1-8-2 with 6 T axle load. These were generally based on the existing Gwalior designs but incorporated many details interchangeable with 2ft 6ins.-gauge locomotives. In 1932, a superheated 4-6-2 appeared, followed by a 2-8-2 superheated of American design in 1948. Amongst the private railways, Martin and Co. used 0-4-2 T locomotive followed by 2-4-2 T and 0-6-2 T classes. McLeod Railways had a large variety of locomotives and during 1950-55, modified ZBs (using saturated steam) were added to its fleet in large numbers. Presently, regular steam traction on the 2ft-gauge is confined only to the Darjeeling Himalayan Railway.

Can one imagine a steam locomotive without fire? Two very unusual designs of steam locomotives also ran in India in the early 20th century. One of these incorporated a steam locomotive without fire, and the other was a monorail, i.e. a locomotive that ran on a single track. While both these unique designs did not

prove to be much of a success, they are excellent examples of innovations in the manufacture of steam locomotives.

A unique design, which did not perform to expectations, was a steam locomotive without fire – the *Fireless Locomotive,* which had a pressure vessel in which steam is collected from a separate boiler. This specially designed locomotive was procured for use in the jute and ordinance factories to eliminate the occurrence of fire, which could have been caused by a spark from a conventional coal-fired steam locomotive. Due to limited steam availability, this locomotive was predominantly used only for shunting operations. It was built in 1953 by Henschell of Germany, and had a maximum speed of 18.5 miles per hour.

In 1907, the first section of an unusual railway based on the Ewing system started in the Patiala state, connecting Bassi with Sirhind. Colonel Bowles, who designed this system, was made the state locomotive engineer and he established the Patiala State Monorail Tramways, about 50 miles in length, from Sirhind to Alampura and Patiala to Bhavanigarh. The track was a single rail on which ran the load-carrying wheels of the train, and a large single wheel, at the end of an outrigger, ran on the road running parallel to the track, to keep the train upright. The German firm of Orenstein and Koppel built the steam locomotives on this unique railway in the year 1907. The double-flange wheels had a wheel arrangement of 0-3-0. One 39in.-diameter flangeless wheel ran on the road. One such train has been kept in working order at the National Rail Museum, New Delhi and offers regular joy rides to tourists.

Next page: Water columns for filling water in steam locomotives were seen everywhere; a WP waiting for its fill.

Indian Hill Railways

One of the most beautiful sights against the backdrop of hills is that of a steam locomotive chugging along, pulling behind it a real train. The Indian Hill Railways provides its travellers with such an experience. Bringing the hills closer to the people is a significant achievement of the Indian Railways. Five railways connecting hill stations to the plains were built in India at the turn of the last century. Three of these link the Himalayan hill stations with the plains in the north; one in the Nilgiri hills in the south links the hill station of Ooty with Mettupalayam; and another links the hill station of Matheran on the Western Ghats with the railway station of Neral in the plains. These five Indian Hill Railways, which pass through a very beautiful and exotic terrain, besides serving their primary purpose of providing transportation, are also a great tourist draw.

B-class steam locomotive hauling a commemorative train on the Neral-Matheran Railway in 2002.

They are wonderful examples of excellence in engineering achieved by the Indian Railway engineers at a time when engineering skills were rather primitive. They have a significant heritage value – an aspect that is now getting increasingly recognised.

Darjeeling Himalayan Railway

Declared a World Heritage Site in 1999, the second railway site in the world to be accorded this coveted status for its outstanding universal value, this railway system is the jewel in the crown of the Indian Railways. It commands universal awe and affection, which is amply reflected in the large and very active membership of the UK-based Darjeeling Himalayan Railway Society. Identified the world over by the cute B-class steam locomotives, built in the UK over 80 years ago, the steam locomotives on this line are its most significant feature, known and loved the world over.

Darjeeling, Darjiling or Dorje-ling means the place of Dorji, the mystic thunderbolt of the Lama religion, and is connected with the cave on the Observatory Hill. According to Hindu mythology, Indra, the god of heavens, jealous of Varuna's (god of wind) dominance of the winds, became furious when one of the winds revolted and established itself in the depths of Mahakal, now known as Observatory Hill. This evil one destroyed all sacred edifices built on the hill and the Lamas were constrained to desert the site. Indra at last threw a thunderbolt called Dorjee and the place got the name Dorje-ling. Until the beginning of the 18th century, the whole area between Sikkim and the plains of Bengal, including Darjeeling and Kalimpong, belonged to the Rajas of Sikkim. In 1706 they lost Kalimpong to the Bhutanese, while control of the remainder was wrested from them by the Gurkhas, who invaded Sikkim in 1780. This annexation by the Gurkhas, however, brought them in conflict with the East India Company. A series of wars were fought by the Raja of Sikkim with the Gurkhas, leading to the defeat of the Gurkhas and the territory was restored to Sikkim. In 1835, the nucleus of what was originally known as British Sikkim was created by the purchase of the sanitarium of Darjeeling and some of its surrounding hills from the Raja of Sikkim for a meagre annual allowance. The ceded tract, about 138 square miles, became a favourite summer retreat for the officials of Bengal and their families. Soon sanitariums and schools for Europeans were opened at Darjeeling, Kurseong and Siliguri.

While Darjeeling was growing, Rowland Macdonald Stephenson was crusading his battle for railway extensions in India. In 1849, he was able to extract favourable conditions including a guarantee of return on the capital. He promoted East India Railway Company and was awarded the construction of an experimental line, running from Howrah to Raniganj. On 15 August, 1854, the first train steamed off from Howrah and by 1855, the complete section was opened to traffic. The route from Calcutta to Darjeeling, then available for those who had the time, money and energy necessary to undertake so formidable a journey, was by rail from Howrah to Sahibganj, a distance of 219 miles; followed by steam ferry across the Ganga to Carragola; thence by bullock cart to the river opposite Dingra Ghat; after crossing which again by bullock cart or a palanquin to Purnea, Kishnganj, Titalya, and Siliguri; whence the ascent commenced via Punkhabari Road, which joined the present Cart Road at Kurseong for onward journey to Darjeeling.

The Darjeeling Steam Tramway Co. with capital fully subscribed in India, was formed in 1880. On 15 September, 1881, the title of the company was changed to Darjeeling Himalayan Railway Co. and this company remained effective until the line was taken over by the Indian Government on 20 October, 1948. Till the time the line was nationalised, it was managed by the agency of Gillanders Arbuthnot & Co. with headquarters at Calcutta, from where it supervised the financial, legal and purchasing interests of the Darjeeling Himalayan Railway and other small railways. The manager and engineer of the line were stationed at Kurseong, while the mechanical superintendent was based at Tindharia, where a workshop for maintaining the steam locomotives, coaches and wagons was established.

Siliguri, the starting point on the line, became the hub of activity as construction on the Darjeeling Himalayan Railway commenced in 1881. After leaving Siliguri, the train's first stop is at Sukna (height 533 ft above sea level), which is 7-1/8 miles from Siliguri (height 400 ft). Between Siliguri and Sukna, the line crosses River Mahanadi on an iron bridge, 700 ft in length (seven spans of 100 ft each); otherwise bridges are few – the only ones are those carrying the line over itself at loops. Sukna is the point whence the trains begin their actual ascent up the hills. At 11-1/2 miles the first spiral or loop in the line is seen. There is a stop for water at 12-3/4 miles. From here the line turns nearly south on to a long spur, where another and somewhat complicated loop occurs. The line then returns north and eastwards, runs for a shorter distance along the road, and gradually passes below

Previous page : Train on the Darjeeling Himalayan Railway—a World Heritage Site.

it, till a third spiral or loop is reached at the 16th milepost, where stands the old Chunabati dak bungalow, a halting place for lunch, which reminds one of the now forgotten days of the tonga.

Tindharia, the first reversing station on the line, brings to mind the success story of the genius that made this fairy-tale ride possible. It was here that the engineer building the line received his first setback. A deep erosion in the hillside made it impossible to employ a gradient within the limits of rail-transport. There seemed to be no alternative but to admit failure, and this he was ready to do when his wife saved the situation. "Darling," she is said to have suggested, "if you can't go ahead, why don't you come back." This brilliant scheme of climbing hills, also known as Z reversing stations, is as simple as it is clever. The train runs forward almost to the edge of the cliff, then backwards at an oblique angle up the hillside, then forward again, this time high enough above the original track to avoid the problem of land erosion. The line thus follows an elongated form of Z. For almost 40 years since the beginning, there were four complete loops and four Z reversing stations. In 1919, the 1 in 20 gradient on the northern descent to Darjeeling was eased by constructing a double spiral, known as the Batasia loop; and the very small radius double loop (then No.2) between Rangtong and Chunabati was replaced by a newly-built Z reversing station to gain 140 ft altitude more easily for uphill trains. Thus, the line now has four loops and five Z reversing stations. At 25-1/2 miles is a halt, one of the numerous for watering. Here, the nature of soil completely changes to rock, known as Sikkim gneiss. A few yards up is the large watercourse known as Pagla Ghora or Mad Torrent. At about the 30th mile, the train passes through rock cuttings. Another mile and Kurseong is reached, which is reminiscent of a typically British climate and countryside.

The railway station is the hub of Kurseong, near which are located most of the shops. At the station, one realises for the first time after seeing the faces of the various Mongoloid races, that one is at the threshold of three close lands. Tibetan women in their long dark robes and necks adorned with charms and beads, pester the travellers to buy the wares of their country for sale in tin boxes; there are the sturdy Nepalese women anxious to carry luggage up or down; and there too are the syce boys with their sturdy mounts and carefree dandywallas with their chair-like devices to carry invalids to the highest point in the town. The train moves out of the station, crossing the bazaar. Below the railway station and towards the north, lie the spacious offices of the Darjeeling Himalayan Railway Co. and later also of Assam Rail Link Project. Now the line is managed from

Next page: Here the train is pushed, not pulled, by an X-class steam locomotive on the Nilgiri Mountain Railway.

S3
SLR
S

LV

Guwahati, where lie the headquarters of the North- East Frontier Railway. Resuming the journey by train from Kurseong, one notices a large building : this is St. Mary's Training College for ecclesiastical students, belonging to the Jesuits. The train also passes along a diversion below the road for about a mile before reaching Tung Station (elevation 5,656 ft). From this point the line runs generally along the Old Cart Road. At 41-1/4 miles, the train reaches the station of Sonada (elevation 6,552 ft). Passing through the bazaar, the traveller is struck by the typical formation of huts and shops, the merry little Bhutia children with their rosy cheeks, and the quaint looks and appearances of their elders. Further ahead of Sonada, the train passes through forests which clothe the hillsides. The train pulls up at Ghoom Station at 7,408 ft elevation – the highest point on DHR. Immediately afterwards, the train proceeds downhill most cautiously, the gradient being a steep 1 in 23 for a short distance. A few minutes more and the train stops at the Darjeeling terminus.

Constructed on the spurs of Himalayan foothills, the loops or spirals are often sites, sometimes breathtaking, as the width of the cars is 3.4 times the rail gauge, and gives passengers the impression that they are over a sheer drop. From early days the more spectacular points were given names, painted on boards, and a government inspector's report of 1888 includes, "I doubt the wisdom of calling the attention of timid passengers to their seeming danger by the exhibition of boards with alarming placards such as Sensation Corner, Agony Point, etc."

A unique feature of this line is that it is virtually an inseparable part of the environment. The line running along one side of the road passes through various villages and towns enroute. The intimacy of the train, the road traffic and the walking public is on display all along the route. At places, the line is so close to the houses on the sides that sometimes one wonders whether the train will actually drift into the lovely little houses! People jumping on and off the running train is a common sight, conveying the closeness of the train and the populace.

Mainly four designs of locomotives have adorned this line. A & B classes built by Sharp Stuart & Co., the C-class Pacific built by the North British Locomotive and the D-class Beyer Garratt built by Beyer Peacock & Co. Ltd, have worked on this line, though the B class has traditionally been the mainstay of the operations. The line is still mainly worked by the B-class 0-4-0 saddle plus well tanks which were built between 1889 and 1927, though diesel locomotives have also been introduced recently by the Indian Railways, despite considerable opposition by heritage lovers of the country. Most of the steam locomotives were built by Sharp

Stewart & Co. in the UK or by their successors, the North British Locomotive Co. in Glasgow. Three locomotives were built in the USA and three were assembled locally, using parts shipped from the UK. This railway is still regarded as the holy grail the world over for the narrow-gauge enthusiast and the name itself commands universal attention for this most famous section of the Indian Railways. The 1 ft-gauge Fairbourne Railway in the Wales has paid the ultimate tribute to the Darjeeling Himalayan Railway by building and operating an exact miniature of the B-class steam locomotive on their system.

Kangra Valley Railway

Kangra valley is the name given to the entire region that lies between the Dhauladhar ranges of the Himalayas in the north and the foothills to the south. It is a

X-37385 of the Nilgiri Mountain Railway seen at its final resting place in the National Rail Museum, New Delhi.

conglomeration of valleys and plateaus. In shape it is roughly an oblong, with a length of about 90 miles and a width of 30 miles through the mountains. To the north they rear skywards – first a low chain of ridges, then a line of peaks averaging between 7,000 and 9,000 ft and directly behind them, mountains rising from 13,000 to 16,000 ft and behind them, the snow-clad mountains. The valley is famous for its natural beauty and ancient Hindu shrines. The Kangra Valley Railway, which runs from Pathankot to Joginder Nagar, proves that the railway engineer can carry out his work in total harmony with the beauty and stateliness of the surroundings. This has been done in the Kangra valley without destroying the grandeur of the mountains, in the process revealing to the tourist, an enchanting fairyland.

The first sod of this 109 miles-long 2 1/2ft-gauge line was cut by the Governor of Punjab on 2 May, 1926, and the 100 mile-line constructed in rough terrain and hostile weather was opened to traffic on 1 April, 1929. The line begins from Pathankot and runs parallel to the road for the first 16 miles. There are 20 crossing stations and seven passenger halts on this line. Ahju at 12 miles is the highest point and there are, in all, 971 bridges and two tunnels on this line. The steel arch bridge over the Reond Nullah and the girder bridge over River Banganga are noteworthy. On this line the curves are relatively easy. For the greater portion of the line, the traveller can gaze continuously at the ever-present panorama of snow-clad ranges and gold green fields, without being swung around every minute on a narrow arc, before his eyes can greet the scenery. As Palanpur is approached, the ever-present background of the snowy ranges draws nearer till the long chain of peaks, 15,000-16,000 ft in height, remain barely 10 miles from the railway. Just beyond Baijnath Piprola station, the line has its most severe gradient, 1 in 19, for 700 ft at mile 58, with approaches of 1 in 31 and 1 in 25. This is the steepest gradient for any adhesion line on the Indian Railways. The line from Baijnath Piprola to Joginder Nagar, the terminus is mostly on 1 in 25 gradient. ZF and ZF1 classes of steam locomotives were originally in use on this line. However, diesel locomotives have now taken over the line completely. Efforts are now being made to restore, with a fixed frequency, at least one steam-hauled service on the line.

Kalka-Simla Railway

The idea of a railway line to Simla dates back to the introduction of railways in India. In the *Delhi Gazette,* a correspondent in November 1847 sketched the

route of a railway to Simla with estimates of the traffic returns, etc. in style. He wrote, "We might then see these cooler regions become the permanent seat of a government, daily invigorated by a temperature adapted to refresh an European constitution, and keep the mental powers in a state of health, beneficial both to the rulers and the ruled."

Survey for a railway line to Simla featured in the administrative reports of Indian Railways year after year. It is interesting to note that the Simla line has been the most surveyed line. The earliest survey was made in 1884, followed by another in 1885. Based on these two surveys, a project report was submitted in 1887 to the Government of India for an adhesion line, 68 miles in length and with a ruling gradient of 1 in 33. After the commencement of the Delhi-Ambala-Kalka line, fresh surveys were made in 1892 and 1893 and two alternative proposals were submitted. During 1894, four more alternate schemes were suggested – two adhesion lines 67–1/4 and 69–3/4 miles long and two rack lines 46–1/4 miles long each. Fresh surveys were again made in 1895 from Kalka to Solan with a view to locate the line either by 1 in 12 rack system or 1 in 25 adhesion system. Lengthy debates followed and finally an adhesion line was chosen in preference to the rack system.

On 29 June, 1898, a contract was signed between the Secretary of State and the Delhi-Ambala-Kalka Railway Co. for construction and working of a 2 ft-gauge line from Kalka to Simla. As per the contract, the rail line was to be built without any pecuniary aid or guarantee from the government; however, the land was provided free of charge. The military authorities were sceptical about the narrow gauge of 2ft chosen for Kalka-Simla Railway. They recommended a standard 2ft 6in.-gauge for hill and light strategic railways. The Government of India yielded to the military requirements and on 15 November, 1901, the contract was revised and 2ft 6in.- gauge was adopted for Kalka-Simla Railway. This meant change of gauge for a portion of the line built in the year 1901. In the beginning, the line was laid with 41-1/4 LB, flat-footed steel rails, 21 ft long, on steel-bearing plates and deodar timber sleepers, nine to a rail. The track was stone ballasted throughout, and fenced only along the Kalka camping ground and through the outskirts of the town of Kalka. The line measuring 59.44 miles from Kalka to Simla was opened for traffic on 9 November, 1903. Thanks to peculiar working conditions, high capital cost coupled with high maintenance cost, Kalka-Simla Railway was allowed to charge higher rates of fare compared to the then prevailing rates for other lines in the plains. By 1904, the Delhi-Ambala-Kalka

Railway faced a serious financial crisis in which a total of Rs 16.5 million was spent. The company represented and the Secretary of State decided to purchase the line from 1 January, 1906.

The landscape along the entire route is magnificent. Flanked by towering hills, the line, like twin threads of silver, clings perilously to the sides of steep cliffs or it ventures boldly over graceful bridges where hundreds of feet below, the little hill streams gush and sparkle in the sunlight. On leaving Kalka, 2,100 ft above mean sea- level, the rail line enters the foothills, commencing its picturesque climb immediately after its departure from Kalka station. The first great difficulty was the huge landslide on the seventh mile of the Cart Road, which extends from the hill summit down to River Khushallia 1,500 ft below. As it was impossible to find a good alignment passing either below or above the slip, construction along the face of the landslide was out of question; the only alternative was to burrow under the hill. A tunnel, nearly a mile long, was constructed in the solid wall behind the disturbed surface strata and is known as the Koti tunnel. The main station, Dharampur, is at a height of 4,900 ft and 20 miles from Kalka. The gradient here is very steep and to achieve flatter gradients required by the railway, the line develops into three picturesque loops at Taksal, Gumman and Dharampur respectively. After leaving Dharampur, the railway gains on the road by taking short cuts and tunnels, so that upto Taradevi, the distance by rail from Kalka is 1/4 mile less than the distance by road, despite railway handicaps. From Taradevi, the rail line goes round Prospect Hill to Jatogh, winding in a series of graceful curves round the Summer Hill and burrows under Inverarm Hill to emerge below the road on the south side of Inverarm, at its 59th mile on to the terminus near the old Dovedell Chambers. At Dagshai, mile 24, the railway line is 5,200 ft above the sea, whence it falls to 4,900 ft at Solan, and to 4,667 ft at Kandaghat (mile 36-1/2), where the final ascent towards Simla begins. Between Dagshai and Solan, the railway pierces the Barogh Hill through a tunnel 3,752 ft long and situated 900 ft below the road.

Throughout its length of 60 miles, the line runs in a continuous succession of reverse curves upto 120 ft radius along valleys and spurs, flanking hills, rising to 6,800 ft above sea-level at the Simla railway station, the steepest gradients being 3 in 100. Kalka-Simla Railway, with its extraordinary feat of engineering skill, more than anything else, contributed to the speedy development of Simla. An interesting feature of the Kalka-Simla Railway is the almost complete absence of girder bridges – multi-arched galleries like ancient Roman aqueducts being the

commonest means of carrying the line over the ravines between the hill spurs. There is only one 60 ft- plate girder span in a pine wood near the old engineer's bungalow at Dharampur, and a steel trestle viaduct which replaced a stone gallery in 1935, in the 869 bridges representing about 3 per cent of the line. The entire section has been built with steep gradient through the Shivalik ranges. Another special feature of Kalka-Simla Railway is that as many as 27 cutovers serve as different gradient crossings. There are 20 intermediate stations and all have crossing facilities. The line also has about 107 tunnels, which, besides being an engineering feat, also generate a lot of interest in the travellers. During the summer months, passenger traffic is heavy whereas in winter, potato traffic keeps the line busy. The line has excellent trains for travellers. Deserving special mention are the luxurious *Shivalik Express,* the rail buses and the luxurious *Shivalik Palace* saloon for tourists. The Kalka-Simla line is characterised by probably the best-maintained hill railway of the country. A recent achievement has been the resurrection of a KC class of steam locomotive, which is stationed at Simla and is available for hauling steam charters to Kethleeghat station and back, a round distance of about 27 miles. A ZF class of steam locomotive is also being resurrected for this line and hopefully the Shivalik hills would buzz once again with the sound of steam locomotives.

Nilgiri Mountain Railway

Coonoor is situated 6,000 feet above the sea at the south-east corner of the Nilgiri plateau, and at the head of the principal pass from the plains. Up the Ghats runs a road (21 miles long) and a rack railway (16¾ miles) from Mettupalaiyam in Coimbatore district. The place was constituted a municipality in 1866. Coonoor remained a terminus for the Nilgiri line for eight years. The extension from Coonoor to Ootacamund was constructed by the Government of India, and the line was opened upto Fernhill on 15 September, 1908, and upto Ootacamund a month later. Rack system was discarded for this extension though the ruling gradient is as severe as 1 in 23. The Ooty terminus was named Udagamandalam, the Tamil word for Ootacamund.

The main attraction/feature of this line is the unique rack system and the equally unique and complicated steam locomotives. To quote from Sir Guilford L. Molesworth's report of 1886: "The locomotive used for working on the Abt system has two distinct functions: first is that of traction by adhesion as in an ordinary locomotive, and the second is that of traction by pinions acting upon the rack bars. The brakes are four in number – two hand brakes acting by friction,

and two acting by preventing the free escape of air from the cylinder and thus using compressed air in retarding the progress of the locomotive. The former are used for shunting whilst the latter for descending on steep gradients. One of the hand brakes acts on the tyres of the wheels in the ordinary manner and the second acts on grooved surfaces of the pinion axle, but can be used in those places where the rack is laid." Even after a hundred years, the brake system on Nilgiri locomotives is as intricate and cumbersome as it was in 1886. The train journey from Madras to Mettupalaiyam (327 miles) then took just over 17 hours and cost Rs 20, first class, and another Rs 20 to cover the remaining 33 miles up the steep hill road to Coonoor and Ootacamund by the 'Nilgiri Carrying Company's Mail and Express Tonga Service', while heavy baggage had to be sent by bullock cart. The only alternative was to hire a pony and arrange for luggage to be taken up by individual baggage carriers, using the shorter but even steeper old road to Coonoor. Now to get to Ootacamund, one can catch the evening *Nilgiri Express* at 9.00 p.m. from Madras Central Station and arrive at Mettupalaiyam at 7.10 a.m. after a 10-hour journey. From here one merely crosses the platform to join the metre-gauge train, which leaves at 7.25 a.m. and reaches Udagamandalam at 11.40 a.m. in less than 15 hours.

The Nilgiri Railway (NMR) is a feat of engineering, unique in the East. The line is metre gauge, practically level for the first 4-1/2 miles, to Kallar at the immediate foot of the hills. As soon as the train leaves Kallar, the rack rail appears and the long climb begins. In the next 12 miles to Coonoor, the line rises 4,363 ft, curving almost continuously as it clings to the hillside, crossing lofty viaducts or tunnels through the hard rock. In this distance, there are nine tunnels, the longest being 317 ft in length. The gradient posts read 1 in 12-1/2 with monotonous consistency. Construction expenses were heavy because in addition to the tunnels, a big bridge over River Bhawani at the foothills was necessary. Besides this large bridge, 26 other bridges, smaller in size, were constructed and heavy expenditure incurred in rock cutting and blasting. "It has been worth it," to quote a South Indian Railway spokesman in 1935. "Those engineers must have been lovers of Nature when they decided on the alignment."

Matheran Light Railway

Abdul Hussain, son of the business tycoon, Sir Adamjee Peerbhoy of Bombay, was a regular visitor to Matheran at the turn of the century. After having obtained a reluctant consent from his father, young Abdul Hussain camped at Neral in

1900 to plan for a narrow-gauge railway line to Matheran. The construction started in 1904 and the 2 ft-gauge line finally opened to traffic in 1907.

Neral, the starting station of this line falls nearly midway on the Mumbai-Pune route of Central Railway. Starting from Neral, the narrow-gauge 2-ft line runs parallel to the main broad-gauge line, leaving the road to the west of Hardal Hill, then turning sharply east. The ascent commences and road and rail almost meet at the end of the third mile near Jummapatti station. They part company, again to meet a mile further, just beyond the steep slope of Bhekra Khud. From here a very interesting portion of the line comes into view. A narrow stretch of level ground terminates in the abrupt rise underlying Mount Barry. To avoid a reversing station, a large horseshoe embankment was constructed. Round this the line runs for a mile in the north direction till it turns back through the only tunnel on the route. 'One Kiss Tunnel' gives a couple time just sufficient for a kiss. We are now halfway through the hills. In the olden days the tiny locomotive may have exhausted all its water. A waterpipe is available and the station is conveniently named 'Water Pipe'. The name continues though the diesel locomotives in use now do not get exhausted and the water pipe has lost its importance; instead a teastall on the platform and a liquor shop few steps up the station serve the passengers on this mid-way point. The line now lies under Mount Barry; to negotiate the rise here, the line zigzags sharply backward and forward twice, passing through two deep cuttings. The line pursues its way more decorously and reaches out more or less straight for Panorama Point, after skirting it, and then returns by Simpson's Tank and terminates close to Matheran Bazaar.

The railway itself is 12-1/2 miles long and of 2feet gauge. The permanent way originally consisted of rail, 30 pounds to a yard, with a ruling gradient of 1 in 20. Speed is limited to 12 mph on straight track but on sharper curves, it is restricted to 5 mph only. Construction of line was done by local labour, though occasional help was sought from the Pioneer Regiments. Heavier ones, weighing 42 pounds to a yard, have since replaced the rails. As a precautionary measure against frequent slides, the line used to be closed during rainy months of July and August till recently, but now passenger services continue even during the rainy months. To commemorate the continuance of trains in the monsoon months of 1982, an M.L.R. locomotive No. 741 (O&K 1767 of 1905) has been installed on a pedestal at Matheran station.

FAIRY-QUEEN EXPRESS
EIR 2

Saga of the Fairy Queen

Fairy Queen — Built in 1855 by Kitson, Thompson & Hewitson of the UK, she is a Guinness record holder for being the oldest working steam locomotive in the world.

Fairy Queen is a name which conjures up childhood images of fairies. She however is a beautiful steam locomotive, that has the distinction of being the oldest preserved steam locomotive in India, and the oldest working steam locomotive in the world. She is a legend of the Indian Railways; a legend that came to life in the year 1997, waking up India to the beauty and nostalgia of steam.

Built in the year 1855, by the British firm of Kitson, Thompson & Hewitson, the *Fairy Queen* is also a *Guinness* world-record holder, a distinction it achieved in January 1998, for becoming the oldest working steam locomotive in the world.

Graceful and majestic, this 26-tonne locomotive with a 2-2-2 wheel arrangement is one of the finest specimens of steam locomotives. This locomotive, which also served the British in transporting troops during the first War of Independence in 1857, remained in active service till 1909, when she was retired and put on a pedestal outside the Howrah railway station. Later, she spent some time at the Railway Training School at Chandausi, as an exhibit for curious students, from where she was taken in 1971 to her supposedly final resting place — the National Rail Museum, New Delhi, as the first exhibit of India's first railway museum.

But fate had something else in store for the sleeping beauty. Her place of rest had to become her place of resurrection. Some 25 years later, February 1996 saw a totally different drama unfolding at the museum: a drama which would gladden the heart of any steam enthusiast. Driven by the burning desire to relive the golden era of steam, a group of steam locomotive experts headed by S.Dhasarathy, the traction man of the railways, was witnessed discussing the capability of the old lady to take to the track once again.

The general opinion being in the affirmative, it was decided to attempt waking the sleeping beauty from slumber and put her back on the rail as a tourism and heritage icon : an icon which would be unparalleled in the world. Thus started the saga of the *Fairy Queen*.

Fairy Queen was not thought of as an end unto itself. Precious as it was for the country, its run was planned for February 1997, as a trigger for many more such heritage runs by different breeds and varieties of steam locomotives. This would have led to a rebirth of the steam era, though on a limited scale for tourism purposes, on lines similar to many countries in the West. A rebirth was required as it had been only two months since the Indian Railways had finished phasing out steam locomotives from its broad-gauge network, thus marking the end of an era.

Fairy Queen had a simple open design and the experts, after assessing the relatively good condition, quickly concluded that a complete overhaul of the lady should be adequate. This was, however, just a shrewd piece of guesswork as the *Queen* had last been in steam almost 87 years ago : a period much greater than the age of the oldest expert! Moreover, the *Queen* had only hand brakes and the experts, out of consideration for safety record of the Indian Railways, considered

it prudent to attach air brakes to avoid what would have resulted in a certain mishap. The *Queen* was then laid on a flat wagon and shipped to the Perambur Workshop of the Southern Railways in Chennai, for resurrection. Perhaps the single most important event in the entire saga was the enthusing up of the Perambur staff, an act delivered to perfection by the legendary mechanical engineer of the Indian Railways, S.Dhasarathy, then Executive Director of the Railway Board.

Simultaneously, work also started on matters of detail, sorting out which, in the absence of precedents, almost became a nightmare for the mandarins of the Rail Bhawan. It was the never- say-die spirit of the then museum administration, supported by personalities like M. Ravindra, the Chairman and D.P. Tripathi, secretary of the Railway Board, which finally saw the idea culminating into reality. The Indian Railways thereafter put the *Fairy Queen* on a pedestal as one of its singular achievements of the 21st century.

The museum authorities planned the historic run as a package of heritage and wildlife. A two-coach train, one of which was to be a chair-car and the other, the support car, was designed to carry 50 discerning passengers on a once-in-a-lifetime two-days, one-night trip to Rajasthan. In the journey to Alwar, the passengers travelled in a luxury coach to the tiger reserve of Sariska, with the night halt at the imposing Sariska palace. A colourful cultural programme and a visit to the tiger sanctuary added to the value of the package. While the package was being finalised, plans for restoring the *Queen* to working condition were also taking shape.

At the Perambur Locomotive Workshop, the *Queen* was stripped, cleaned, repaired, refitted and brake equipment added. After successful trials, the *Queen* travelled back to New Delhi, piggyback on a flat wagon, where it was received with great fanfare in December 1996. Meanwhile, the museum authorities had finalised the package, and started selling it even before the trials were over.

The historic run of the *Fairy Queen,* scheduled for February 1997, was expecting patronage from international steam enthusiasts. The package, based on the conventional fully-distributed costing approach of the Government of India, was priced at a hefty US $ 500 per person. Unfortunately, the run had to be aborted

FAIRY-QUEEN EXPRESS
EIR 22

as only four tickets could be sold, and with economics outweighing all other considerations, the train was not permitted to run with only four paying passengers on board. The *Fairy Queen* returned to the National Rail Museum, where it was received by the tearful museum director and an equally distraught Ojikutu, a German diplomat who bought the first ticket. But fate had something else in store. This setback could not diminish the indefatigable spirit behind the venture. The reasons for the failure were assessed and it transpired that the very small lead time for marketing coupled with the high tariff was responsible for the debacle. It was decided to run the *Fairy Queen* once again, and this time not as a one off run, but five runs covering the entire tourist season.

A progressive Railway Board okayed the proposal. Finalising the schedule of the *Queen* for the next season and announcing it immediately during the International Tourism Bourse in Berlin, in March 1997, initiated corrective measures in this direction. Before the scheduled commercial run, a commemorative run was planned for 13 August, 1997. Just before the run, a technical complication with the plain bearings of the driving axle cropped up. A bold decision resolved the technical problem and saw the train through the commemorative run, which was extensively covered by the world media.

The next run scheduled for 18 October, 1997, again had only four buyers and the system again developed cold feet. The situation this time was different with M. Ravindra, Chairman of the Indian Railway Board, permitting the running of the train and lowering of the rates to US $ 150 – indeed a saleable rate. The rest is history. Success did not elude the believers this time. The five runs were a roaring success, both technologically and commercially, proving the viability of such ventures. The train achieved an overall 60 per cent occupancy and the achievement was hailed worldwide.

On 13 January, 1998, the train made history by setting a *Guinness* world record for being the oldest working steam locomotive in the world. This was followed by national recognition in the form of a National Tourism Award for the most innovative tourism product for the year 1997-98.

The major significance of this event lay in the fact that it marked the return of steam to India. However, it was accepted that steam would be utilised only

partially and that too only on short sections for tourism purposes, and may never again haul mainline passenger and freight trains on the national railway system. The *Fairy Queen* continues to chug along every season, hauling the specially designed tourist train.

7200
WP

Indian Steam Heritage Preservation Efforts

National Rail Museum: Steam locomotive preservation efforts were started in India in right earnest in the early 1960s, with the appointment of Mike Satow as advisor to the Ministry of Railways for heritage preservation efforts. Serious work, however, began in 1971 with the laying of the foundation stone for the country's first rail museum, the National Rail Museum, at New Delhi. Six years later, the museum was inaugurated. This museum, which was initially conceived as a National Transport Museum meant to preserve rail, road, air and marine heritage, ultimately, however, evolved only as a Rail Museum with primary focus on the steam heritage of the country. Mike Satow, who is rightly the father of the museum, made the initial collection. The museum now houses a wide variety of steam locomotives, including the broad-gauge *Beyer Garrat*, the heaviest locomotive ever to have ruled the track in India, *F-734*, the first indigenously built metre-gauge steam locomotive of the country, *CS- 775,* the

WT 594 steam locomotive of 2 ft 6 in.- gauge. Built in 1925 by W.G. Bagnall, UK; preserved at National Rail Museum, New Delhi.

lightest locomotive of India, *Ramgotty*, the second oldest steam locomotive in India, the manly yet glamorous bullet-nosed *WP*, once the pride of the Indian Railways fleet and many others. The jewel in the crown, of course, is the *Fairy Queen*. The preserved locomotives represent a very fine cross-section of the iron horses to have traversed the Indian sub-continent in the past.

The museum, spread over 10 acres in the posh Diplomatic Enclave of Chanakyapuri, continues to grow as custodian of the heritage of Indian Railways. Addition of relatively new rolling stock, like diesel and electric locomotives, has given a new dimension to the heritage preservation efforts. While a lot has been done in the setting up and expansion of the museum at Delhi, a sane realisation that the Indian Railways is too large for one museum alone to do justice to heritage preservation efforts, has also dawned. Formation of regional rail museums and mini rail museums to adequately address regional and sectoral

ST-707 broad- gauge steam locomotive, built in 1904 by North British Locomotive Works, Glasgow, for the North-Western Railway; preserved at the National Rail Museum, New Delhi.

XT-36863 broad-gauge steam locomotive built in 1935 by Fried Krupp and Co., Berlin for Bombay-Baroda and Central India Railway; preserved at National Rail Museum, New Delhi.

heritage aspirations has also been considered necessary. Accordingly, mini rail museums have been set up at Mysore, Nagpur and Ghum and a relatively larger regional rail museum has been established at Chennai. There is a plan to set up another regional rail museum at Varanasi to cover the eastern region. All these steps are expected to give the desired thrust to steam heritage preservation activities in the country.

Indian Steam Railway Society

Shared passion for steam locomotives and the desire to make a personal contribution saw the coming together of a core group of hardcore steam fanatics. On 23 October, 1999, these enthusiasts met and decided to form a non-profit organisation to further the cause of steam locomotives in India. And the Indian Steam Railway Society was born. The founder members included Mark Tully, an old India-hand and formerly associated with BBC, the renowned travel-writer Bill Aitken, and eminent railwaymen like R.C.Sethi, Harshvardhan, P.J. Singh, Ravindra Gupta and the author. The society was formed with the aim of promoting interest in and sharing the knowledge of the Indian steam heritage and current developments amongst steam enthusiasts by bringing them to a common platform, to stimulate public affection through intelligent exposure of the steam railways, to create awareness of the Indian Steam Railway Works and to highlight the need for their preservation in the interest of science and technology, facilitating transport and tourism. In pursuance of these objectives, the society publishes a quarterly newsletter, organises regular meetings, debates, seminars, competitions and quiz competitons for the members and the public to exchange and further disseminate information on steam railways. It also assists in preservation and revival of the heritage of steam railways in India and promotes steam heritage tourism, advises and, if needed, assists the Government of India in the formulation and implementation of policies related to running steam on Indian Railways and in the process, preserve the working steam heritage for future generations to come.

Other Revival Efforts

The successful run of the *Fairy Queen* and the fame it achieved propelled various railways and railwaymen into action. Steam locomotives started getting

recognised. Nostalgia surfaced. Passion erupted. Discussions on matters relating to steam began, even if for its snob value in elite railway circles. Various zonal railways started competing in organising steam locomotive runs with great fanfare. The first on the bandwagon was the Eastern Railway, which on 19 September, 1999, ran a 77-year- old HGS-class steam locomotive, number 26761, hauling a four-coach train, called the *Millennium Express*, from Howrah to Tribeni, a distance of approximately 32 miles. The train took exactly the same route, which the first train on the East Indian Railway took on 14 August, 1854. Moving at a restricted speed of 17 mph, the return journey consumed the better part of the day. This run was a roaring success.

The North-Frontier Railway surprised everyone by taking major steps in this direction, perhaps due to the phenomenal push given by Bisht, the then General Manager, who belonged to the breed of individuals who get a rush of adrenaline while discussing steam locomotives. This railway revived a AWD-class broad-gauge steam locomotive in early 2000 and running it commercially between Guwahati and Pandu as a tourist package, titled 'Brahmputra' by steam. Another package launched by this railway was the Jatinga Steam Safari utilising a YG-class metre-gauge steam locomotive.

Not to be outdone was the Central Railway, which launched a steam train on 16 April, 2002, thereby heralding the advent of the 150th year of the Indian Railways. Hauled by the bullet-nosed WP-class steam locomotive, this train ran from Mumbai Victoria Terminus to Thane and was a re-creation of the first train run on 16 April, 1853. Unprecedented public response to this run spoke volumes about the affection that steam locomotives command amongst the populace. Crowds lined the sides of the track all along the route.

The lovely hill station of Matheran on the Western Ghats buzzed with the shrill whistle of a steam locomotive, when in April 2002, a B-class steam locomotive, specially brought from the Darjeeling line and restored at the Parel Workshops in Mumbai, hauled a train up the Ghats from Neral to Matheran as a precursor of things to come. Nitish Kumar, the Minister of Railways, accompanied by S.Dhasarathy, member of Railway Board, undertook this unique journey.

Following page : Patiala State Monorail train in action at the National Rail Museum, New Delhi. Built in 1907 by Ornstein & Koppel, Berlin, for the princely state of Patiala.

A major push to steam heritage was given in August 2002 with the inauguration of the Heritage Steam Shed at Rewari. Designed to house a total of 12 steam

ST Class broad gauge steam locomotive built for the North Western Railway at the National Rail Museum, New Delhi.

2' 6" gauge BK class steam locomotive in immaculately preserved condition at the Rambagh Palace in Jaipur.

PSMT
PSMT

locomotives, comprising an equal mix of broad-gauge and metre-gauge steam locomotives, this shed is destined to become the central pivot for steam operations in India. Being developed as a centre for steam locomotives, this shed will house working steam locomotives, steam cranes, steam miniatures and, most importantly, a centre for preservation of steam locomotive skills. Another major milestone has been the manufacture of new steam locomotives, an activity which has been started after a gap of over 30 years. The Locomotive Workshop at Tiruchirapalli on the Southern Railway is busy manufacturing B-class steam locomotives for the Darjeeling line and is also gearing up for manufacturing the X class for the Nilgiri line.

Facing page: FMA 37302 metre-gauge steam locomotive, built in 1888 by DUBS and Co., Glasgow, for the Southern Maharatta Railway; preserved at the National Rail Museum, New Delhi.

Steam heritage tourism has come to stay, though it is yet to reach a crescendo. Seeing the favourable trends in various countries of the world, one is inclined to reasonably believe that the growth of the steam heritage tourism segment would continue unabated. An increasing number of steam locomotives would be brought back from the graveyards and resurrected, new steam locomotives would be manufactured, though on a limited scale, and the sector would witness

Below : B-26 broad gauge, steam locomotive, built in 1870 by Sharp Stewart and Co., Manchester; preserved at the National Rail Museum, New Delhi.

HUBLI

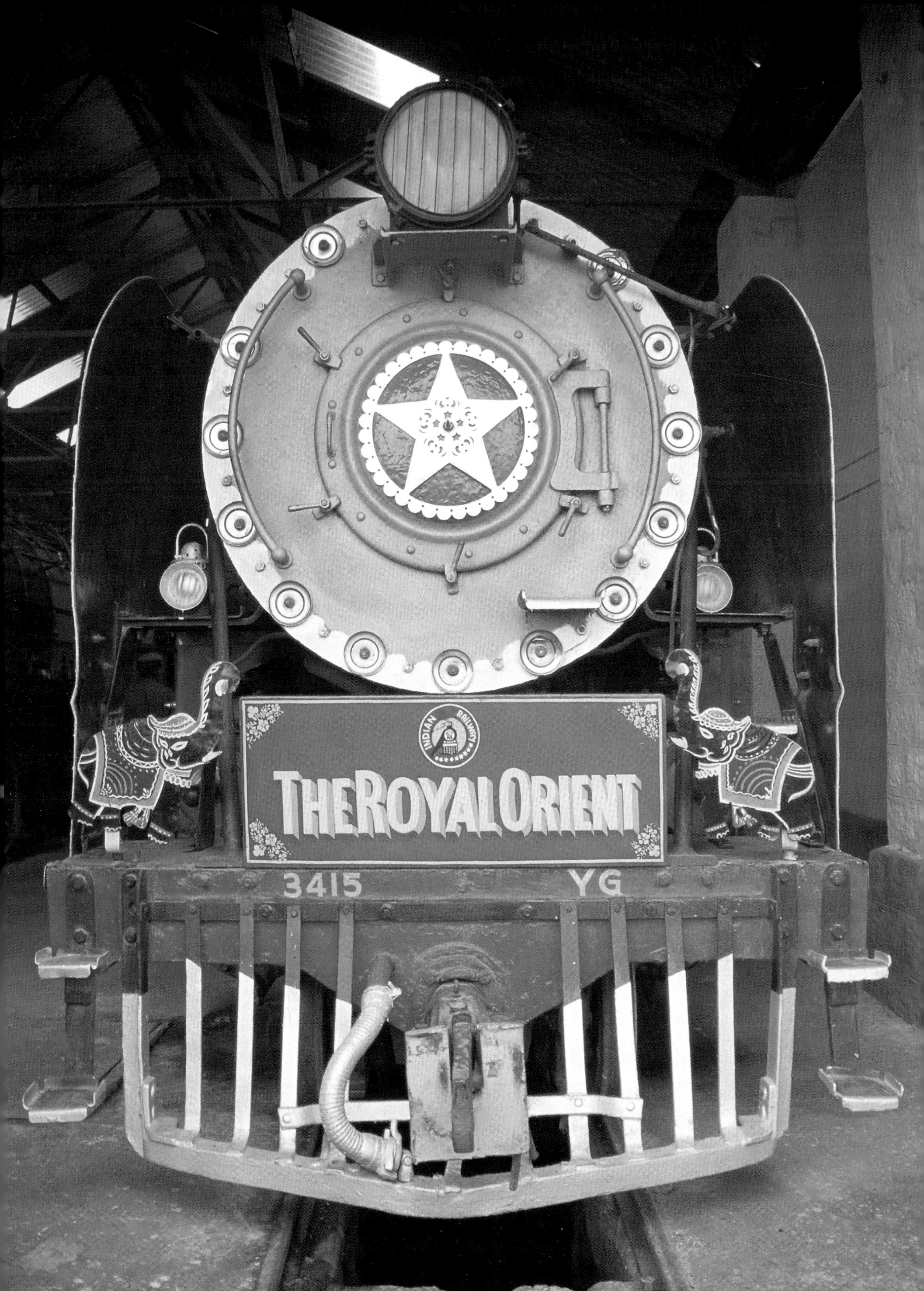
INDIAN RAILWAY
THEROYALORIENT
3415
YG

increasing voluntary involvement. It is hoped that in the foreseeable future, one would see the formation of an international, non-governmental organisation which will co-ordinate the efforts being made in various countries and give them a fillip by generating global affection and awareness for these black beauties.

In India one may see a comeback of steam locomotives, of course, in a limited way, on the five hill railways, possibly on the *Golden Triangle*, on the famous *Palace on Wheels* between Delhi and Jaipur, on the Vadodara narrow-gauge section, somewhere in Kerala and on the historic Mumbai-Thane route. These runs would help preserve this glorious heritage and give an impetus to tourism by carving out a greater role for the niche steam heritage tourism segment. We may also witness a trend similar to that prevailing in the UK, where voluntary effort would be forthcoming in building, maintaining and operating steam locomotives in various sections in the country.

Steam is definitely back, and that too full throttle!

Facing page: YG-class metre- gauge steam locomotive decked up for Royal Orient luxury train at the Rewari Steam Centre.

1598-HG/C broad-gauge steam locomotive, built in 1709 by Vulcan Foundry, UK, for the North-Western Railway; preserved at the National Rail Museum, New Delhi.

Glossary of Terms

Gauge – Width of the rail track

Broad gauge 5'6" width of the rail track

Standard Gauge At 4' 8 &1/2", predominant gauge in UK, Europe & USA

Meter Gauge 1 meter gauge

Narrow Gauge 2' and 2'6" gauges, normally utilized in hills

Aquaduct – Bridging structure for crossing a gorge

Embankment – A track formation raised above ground level

Valve Gear An arrangement of valves for timely injection of steam in the cylinders. These are normally of two designs Walschearts and Caprotti

Tender – A wheeled attachment to a steam locomotive for carrying coal and water

Tank type locomotive – A steam locomotive without tender, having tanks mounted on the boiler itself for carrying water

Wheel Arrangement – Normally depicted as a-b-c, where a is the number of leading wheels, b driving wheels and c trailing wheels of a steam locomotive

Firebox – Where the fire burns for heating the water to generate steam

BESA, IRS – Standardised designs of steam locomotives, pre and postworld war 1

Garratt – 4 cylinder compound steam locomotives used for hauling very heavy loads

Ewing System – A design system evolved for the monorail train in Patiala State Railway

Flangeless wheel – Normally a train wheel has flanges to prevent it from dropping down from the rail. A flangeless wheel does not have a flange to have a higher adhesion

Gradient – Expressed as a ratio of height gained to distance travelled

Ruling Gradient – The maximum gradient on any railway section

Rack & Pinion system – Utilised on the Nilgiri Mountain Railway in India. It provides additional traction and prevents the train from rolling down

Adhesion type locomotive – Locomotives relying on the adhesion between rail and wheel for pulling train

Boiler – Where water is boiled into steam. It has smoke, flue and element tubes inside

Cylinder Arrangement – Normally there are two cylinders and these are either inside or outside the locomotive main frame.

Standard Locomotive Terminology – Normally the first letter denotes the gauge and the second the type of service for which the locomotive has been built. But there are a large number of exceptions, which do not relate at all to this explanation.

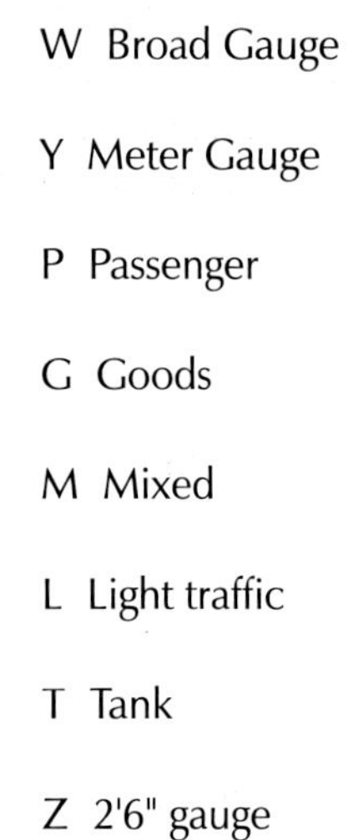

W Broad Gauge

Y Meter Gauge

P Passenger

G Goods

M Mixed

L Light traffic

T Tank

Z 2'6" gauge

N 2' gauge

Steam Locomotive Holding in India in 1977

It is interesting to note that the steam locomotive holding in India was over 8000 as late as 1977. It is precisely for this reason why India became the favourite haunt of the international steam enthusiast after the last of the steam died on the main lines in the west. Substantial steam operations continued in India till the early 1990's. The railway wise holding of steam locomotives towards the middle of 1977 is given below for better appreciation of massive steam operations which were everyday matter till very recently .

Broad Gauge

CR	ER	NR	NFR	SR	SCR	SER	WR	NER	Total
712	1017	1057	77	389	499	586	501	-	4838

Meter Gauge

CR	ER	NR	NFR	SR	SCR	SER	WR	NER	Total
39		317	372	516	347		633	811	3035

Narrow Gauge (2'6")

CR	ER	NR	NFR	SR	SCR	SER	WR	NER	Total
64	11	16		7		106	89		293

Narrow gauge (2')

CR	ER	NR	NFR	SR	SCR	SER	WR	NER	Total
29	4		24						57

Total Steam Holding

CR	ER	NR	NFR	SR	SCR	SER	WR	NER	Total
844	1032	1390	473	912	846	692	1223	811	8223

Type-wise Allocation of Steam Locomotives in India in 1977

Broad gauge

Class	CR	ER	NR	NFR	SR	SCR	SER	WR	NER	Total
WP	104	132	157	14	63	113	73	93		749
WG	439	379	455	40	142	306	395	266		2422
WL	2		63		31			8		104
WT		14			9	6				29
AWC					31					31
XB					9	15				24
XD					68	49	44			161
XE1					16			13		29
HPS					20					20
A/CWD	92	220	287	23				94		716
AWE	15							22		37
H	27							5		32
WM		37	11			10	5			63
AWC		14								14
XC		23								23
XE		27								27
HPS1		36								36
HGS		72								72
HT		33								33
SG		30	17							47
XA	33		5							38
HS							35			35
HSM							34			34
XT			9							9
HPS2			49							49
WW			4							4
Total	**712**	**1017**	**1057**	**77**	**389**	**499**	**586**	**501**		**4838**

Meter gauge

Class	CR	ER	NR	NFR	SR	SCR	SER	WR	NER	Total
YP	9		101	98	200	72		155	234	869
YG	24		168	16	178	151		161	296	994
YL	6		13		46			74	130	269
WD				116	34			75	76	301
YD					17	71		45		133
B					7			6	1	14
HPS					4				10	14
ST					18					18
X					12					12

YM						12				12
YB			2			33		37	39	111
YK						5				5
GS						3				3
YF			6	46				4		56
G1-3			4							4
HG			9							9
HP(New)			10							10
PT			4							4
P									29	4
T									6	6
B1									3	3
B2									11	11
BR									3	3
D2									1	1
G									10	10
GR									5	5
MJ									6	6
NS									3	3
NIS									4	4
OJ									2	2
PGS									2	2
PTS									2	2
TJ									1	1
GX									2	2
GE									44	
Total	**39**		**317**	**372**	**516**	**347**		**633**	**811**	**3035**
Class	**CR**	**ER**	**NR**	**NFR**	**SR**	**SCR**	**SER**	**WR**	**NER**	**Total**

Narrow gauge (2')

Class	**CR**	**ER**	**NR**	**NFR**	**SR**	**SCR**	**SER**	**WR**	**NER**	**Total**
ML	4									4
ND	3									3
NH/2	2									2
NH/3	4									4
NH/4	4									4
NH/5	5									5
NM	8									8
B				24						24
CS		4								4
Total	**30**	**4**		**24**						**58**

Narrow gauge (2'6")

Class	**CR**	**ER**	**NR**	**NFR**	**SR**	**SCR**	**SER**	**WR**	**NER**	**Total**
ZP	3				2					5
ZD	6									6

Class	CR	ER	NR	NFR	SR	SCR	SER	WR	NER	Total
ZE	12		5				44			61
ZF			9							9
KC			2							2
ZA	5									5
ES					3					3
LTS					2					2
A/1	3									3
B/1	7									7
BS	3						22			25
D/1	3									3
F	13									13
G	9									9
ZB								43		43
B								4		4
C								3		3
K								1		1
T								4		4
P								3		3
W								21		21
W/1								4		4
WT								6		6
BC							5			5
CC							21			21
PL							8			8
BAGNALL		7					4			11
DELTA		4					2			6
Total	**64**	**11**	**16**		**7**		**106**	**89**		**293**

Salient details of some major classes of Steam Locomotives of the Indian Railways

Gauge	Class	Wheel Arrangement	Service
Broad gauge	XA	4-6-2	Branch Passenger
	XB	4-6-2	Light Passenger
	XC	4-6-2	Heavy Passenger
	XD	2-8-2	Light Goods
	XE	2-8-2	Heavy Goods
	XF	0-8-0	Light Shunting
	XG	0-8-0	Heavy Shunting
	XP	4-6-2	Experimental Passenger
	XT	0-4-2T	Light Tank
	WP	4-6-2	Passenger
	WG	2-8-4	Goods
	CWD	2-8-2	Mixed
	AWD	2-8-2	Mixed
	WM	2-6-4T	Mixed Tank type
	WT	2-8-4T	Goods Tank type
	WL	4-6-2	Light Passenger
	HSG	2-8-0 + 0-8-2	Garratt Heavy duty Goods
	N/NM	4-8-0 + 0-8-4	Garratt Heavy duty Goods
	P	4-8-2 + 2-8-4	Garratt Heavy duty Goods
	HPS	4-6-0	Heavy Passenger
	HGS	2-8-0	Heavy Goods
Meter Gauge	WD	2-8-2	Heavy Goods
	YP	4-6-2	Passenger
	YG	2-8-2	Goods
	YL	2-6-2	Light Services
	YM	2-6-4T	Mixed Tank
	YB	4-6-2	Passenger
	YC	4-6-2	Heavy Passenger
	YD	2-8-2	Goods
	YF	0-6-2	Branch Lines
	YK	2-6-0	Branch Lines
	YT	0-4-2T	Light Tank
	X	0-8-2T	Rack Railway at Ooty
2'6" Gauge	ZA	2-6-2	
	ZB	2-6-2	
	ZC	2-8-2	
	ZD	4-6-2	

Gauge	Class	Wheel Arrangement	Service
	ZE	2-8-2	
	ZF	2-6-2T	
	B	2-8-2T	
	CC	4-6-2	
	PL	0-6-4T	
	RD	2-6-2	
2' Gauge	CS	2-4-0T	
	Bagnall	0-6-4T	
	Delta	2-6-2T	
	B	0-4-0ST	
	C	4-6-2	